The Treatment and Handling of Wastes

Technology in the Third Millennium

The Treatment and Handling of Wastes

Edited by
A.D. Bradshaw,
Sir Richard Southwood and
Sir Frederick Warner

Published by Chapman & Hall for The Royal Society

The Royal Society

 CHAPMAN & HALL
London · New York · Tokyo · Melbourne · Madras

Published by Chapman & Hall, 2–6 Boundary Row, London SE1 8HN

Chapman & Hall, 2–6 Boundary Row, London SE1 8HN, UK

Van Nostrand Reinhold Inc., 115 5th Avenue, New York NY10003, USA

Chapman & Hall Japan, Thomson Publishing Japan, Hirakawacho Nemoto Building, 7F, 1-7-11 Hirakawa-cho, Chiyoda-ku, Tokyo 102, Japan

Chapman & Hall Australia, Thomas Nelson Australia, 102 Dodds Street, South Melbourne, Victoria 3205, Australia

Chapman & Hall India, R. Seshadri, 32 Second Main Road, CIT East, Madras 600 035, India

First edition 1992

© 1992 The Royal Society and the authors of individual papers

Typeset in 10/12 Times by Keyboard Services, Luton, Beds

Printed in Great Britain by St Edmundsbury Press, Bury St Edmunds, Suffolk

ISBN 0 412 39390 5   0 442 31461 2 (USA)

A catalogue record for this book is available from the British Library

Library of Congress Cataloging-in-Publication data

The Treatment and Handling of Wastes / edited by A.D. Bradshaw,
    Sir Richard Southwood, and Sir Frederick Warner. – 1st ed.
        p.   cm. – (Technology in the third millenium : 1)
    Includes bibliographical references and index.
    ISBN 0–442–31461–2
    1. Refuse and refuse disposal.   2. Refuse and refuse disposal –
Environmental aspects.   3. Pollution. 4. Man – Influence on nature.
I. Bradshaw, A.D. (Anthony David).   II. Southwood, Richard, Sir.
III. Warner, Frederick, Sir, 1910–   .   IV. Series.
TD791.T69   1991
    628.4–dc20   91–2869
CIP

Contents

Part Two
Transformation and Re-use

Part Three
Dispersal

Part Four
Degradation

Part Six
Conclusions

Contributors

Professor G.K. Anderson
Department of Civil and Geotechnical Engineering
University of Newcastle upon Tyne
NE1 7RU
UK

M. Backman
TEM
University of Lund
Box 725
S220–07 Lund
Sweden

Professor A.D. Bradshaw, F.R.S.
Department of Environmental and Evolutionary Biology
The University
PO Box 147
Liverpool
L69 3BX
UK

Dr R.P. Bringer
Staff Vice-President, 3M
Environmental Engineering and Pollution Control
Building 21–2W–07
900 Bush Avenue, PO Box 33331
St Pauls
Minnesota 55133–3331
USA

Professor A.T. Bull
Department of Biology
The University
Canterbury
Kent
CT2 7NJ
UK

D.J.V. Campbell
AEA Technology
Harwell
Oxfordshire
OX11 0RA
UK

Dr P.F. Chester
National Power Technology and Environment Centre
Kelvin Avenue
Leatherhead
Surrey
KT22 7SE
UK

Mr R.A. Frosch
General Motors Research Laboratories
Warren
Michigan 48090–9055
USA

N.E. Gallopoulos
General Motors Research Laboratories
Warren
Michigan 48090–9055
USA

J.E. Hall
Water Research Centre
Medmenham
PO Box 16
Marlow
Buckinghamshire
SL7 2HD
UK

Dr W.D. Halstead
Corporate Planning Department
National Power PLC
Sudbury House
15 Newgate Street
London
EC1A 7AU
UK

Professor J.L. Knill
NERC
Polaris House
North Star Avenue
Swindon
SN2 1EU
UK

Dr L. Kramer
DGXI EEC
Rue de la Loi 200
B1049 Brussels
Belgium

T. Lindhqvist
TEM
University of Lund
Box 725
S220–07 Lund
Sweden

P.T. McInerney
UK NIREX Ltd.
Curie Avenue
Didcot
Oxfordshire
OX11 0RH
UK

R.J. Pentreath
MAFF
Fisheries Laboratories
Directorate of Fisheries Research
Pakefield Road
Lowestoft
Suffolk
NR33 0HU
UK

M.B. Pescod, OBE
Department of Civil and Geotechnical Engineering
University of Newcastle upon Tyne
NE1 7RU
UK

H.G. Pullen
Cleanaway Ltd.
Technical Services
Airborne Close
Arterial Road
Leigh-on-Sea
Essex
UK

Dr J. Rae
AEA Technology
Harwell
Oxfordshire
OX11 0RA
UK

Professor L.E.J. Roberts, C.B.E., F.R.S.
Environmental Risk Assessment Unit
School of Environmental Sciences
University of East Anglia
Norwich
NR4 7TJ
UK

Dr F.B. Smith
Meteorological Office
London Road
Bracknell
Berkshire
RG12 2SZ
UK

Sir Richard Southwood, F.R.S.
Department of Zoology
South Parks Road
Oxford
OX1 3PS
UK

Professor C.W. Suckling, C.B.E., F.R.S.
Willoway
Shoppenhangars Road
Maidenhead
Berkshire
SL6 2QA
UK

R.K. Turner
Environmental Appraisal Group
School of Environmental Sciences
University of East Anglia
Norwich
NR4 7TJ
UK

Sir Frederick Warner, F.Eng, F.R.S.
Cellar House
11 Spring Chase
Brightlingsea
Colchester
Essex
CO7 0JR
UK

Dr D.S. Woodhead
MAFF
Fisheries Laboratories
Directorate of Fisheries Research
Pakefield Road
Lowestoft
Suffolk
NR33 0HU
UK

Part One

The Nature of the Problem

Chairman's introduction

Sir Frederick Warner

This book is directed towards the future. But it necessarily has to begin from where we are. It gives some leads to the scientific and technical possibilities for environmental protection. This can help decision-makers to develop a regulatory framework that is sensitive, progressive and cost effective.

The original meeting was entitled 'Approaches to the handling and treatment of wastes'. This did not, however, exclude discussion on how to prevent wastes. There are two important first principles for all processes: the conservation of mass, and the conservation of energy. These focus attention on the starting point of any individual process, the information that exists by which mass and energy balances for the process can be constructed. These are basic tools of trade in the discipline of chemical engineering for the design of chemical processes. Such processes can usually be accomplished by different routes which can be characterized by flow sheets. These quickly show that pollution and rejects of material or energy to air, water or earth are interlinked. The scrubbing of gases to clean them can produce polluted water and solid wastes. Cleaning polluted water can produce gases and solids, themselves pollutants.

Conservation of matter and energy is easy to understand and follow in closed systems. It is more difficult to apply where wastes are subject to external methods of disposal. Where they are discharged into the wider environment, their routes of dispersion and chemical transformation can be particularly difficult to specify. There are considerable problems, for instance, in understanding gas reactions that lead to ozone destruction or predicting bio-accumulation in food chains. An outstanding example of the latter was the accumulation of methylmercury in fish at Minimata Bay where the rate of renewal of seawater was low. Radionuclides, whether they arise from reprocessing of nuclear fuel, bomb tests or accidents such as Chernobyl, can be used as tracers to show some of the complexities of pathways in both the marine environment and in soil, where adsorption, ion exchange, displacement by mass action, and removal through plants and animals occur.

All this shows that an understanding of the effects of wastes is going to be difficult. It is crucial that their biological effects are properly understood (discussed by Bradshaw). It is relatively easy to see what these are likely to be when materials are immobile. But when materials disperse readily through the environment, the situation is much more difficult. It is easy to fall into the trap of thinking that dispersal and dilution provide a simple answer to disposal problems. But materials which do not degrade into substances which are innocuous are liable to accumulate somewhere in the biosphere and cause serious problems. It may be some time, as in the case of organo-chlorine pesticides (Ratcliffe, 1970), before the effects become apparent. It may also be that apparently innocuous degradation products, such as carbon dioxide and nitrates, have serious indirect effects.

Here the biological problems are immediately thrown into sharp relief, because of our still incomplete knowledge of the balance of biological systems, and the difficulties that remain in quantifying flows of either energy or materials such as carbon or nitrogen. The increased interest in the bio-geochemical cycles of both these elements, particularly carbon, in relation to climate change, has shown that there are considerable unfilled gaps in our knowledge.

The object of design is to find the most economical solution. This is not confined to the balance between the cost of inputs and the values realised in sales in a limited sense. The inputs have to take into account the short and long term environmental costs, including the costs related to occupational health effects and the wider social costs shown up by risk assessment. These have been discussed in the Royal Society Report 'Risk assessment', particularly in Section 6.6 on Risk Management.

When all costs have been taken into account, the whole balance of judgement on the way a waste should be handled, or on whether a particular product should be produced at all, may be radically altered. Completely different conclusions as to what is acceptable may be reached (discussed by Backman & Lindhqvist). The idea that each production activity, or part of a productivity activity, must be assessed is simple, but radical. It has substantial ramifications, not the least being that the product itself must be assessed for its environmental impact. Because, eventually, the product is used up, or worn out and discarded, it too becomes a waste.

If anything is ever to be achieved, concepts about environmental management have to be translated into action. Since human beings are both selfish and not good at understanding the long-term effects of their actions, for wastes this action means the establishment of regulations to specify what is, and what is not, to be permitted, for the common good. An indication of the developing European view is given by Kramer. There will no doubt be discussion as to how far any regulation system should go, but the potential seriousness of the pollution problems raised by present day waste disposal means that tighter regulation is inevitable.

These considerations are of immediate interest in the U.K., in relation to the proposals of the Environmental Protection Bill, from which some deductions can be made of the underlying methodology. Part I is based on integrated pollution control through central inspectors, with air pollution from prescribed processes designated for local control by municipal authorities. This system is proposed in the 5th Report of the Royal Commission on Environmental Pollution (the comparison was made between sheriffs and deputies in Westerns, a film-form popular then, and even now in television) (RCEP 1975).

Part I Section 3 of the new Bill gives power to make regulations establishing standards, aims or requirements. In relation to standards, limits may be set on concentrations, amounts or rate of release; and methods of measurement can be prescribed. This should allow emphasis on mass balance for longer term effects compared with problems of concentrations in acute episodes. It also provides for limits to be set on the total discharged (and maximum rate) into the environment of the U.K.

A concept is then introduced of quotas. An analogy can be made with the benefits arising from the enclosure of common land which allowed the U.K. agricultural revolution in the 18th century. Environmentalists have often referred to 'the tragedy of the commons', in which uncontrolled use by all leads to over exploitation or neglect by all (Hardin, 1977). The new proposals would meet this problem by giving Government the power to restrict access for industrial plants to those authorized, to allocate shares in the amounts that an industry may release, but to retain the power to plan progressive reductions in the standard set following progressive improvement in quality objectives. In comparison with land enclosure, no freeholds are to be granted, only conditional licences. Whether these will confer any kind of title capable of being transferred at the will of the licensee has yet to appear in the legislation.

In authorizing discharges with these conditions, the aims are to comply with local, European Community, and international obligations, by using the best available techniques, not entailing excessive cost, to reduce or render harmless any release. The best techniques include (in addition to technical means) the number, qualification, training and supervision of persons employed in the process, and the design, construction, layout and maintenance of the buildings in which it is done.

Underlying all the discussion in this book is the concept of harm. In the Environment Protection Bill, it is defined as meaning 'in relation to living organisms, harm to their health and, in the case of man, includes offence to any of his senses or harm to his property'. In respect of man, for whom most information is available, no great advance has been made since the late Sir Edward Pochin wrote about the 'Problems involved in developing an index of harm' (Pochin, 1977). It is certainly not always easy to determine when harm is actually occurring. For living organisms, the modelling of population dynamics shows that natural fluctuations are large. As a result, the impact of environmental pollution is difficult to assess except where it is massive. For the North Sea, for instance, even sewage sludge from London dumped in the Barrow Deep, and direct discharge of partly treated sewage, seem to cause no harm if the productivity of the North Sea is the measure.

Yet biological indicators, in particular cases, can be reliable evidence of harm: the state of the Thames Estuary 20 years ago is an example. The lack of oxygen caused mainly by the discharge of partly treated sewage made life impossible for fish, but favoured two populations – anaerobic bacteria, and, in the benthos, tubificids, with densities of 10^6 m^{-2}. When the pollution was overcome in the mid-70s by better sewage treatment, tubificids came down to the more normal level of about $ca.10^4$ m^{-2}, and nearly 100 varieties, of fish returned (Andrews, 1984). This kind of evidence is not available in an environment which is more difficult to characterize by a model than an estuary. It has been difficult to link, for example, epidemics of dinoflagellates in lakes and reservoirs with specific pollutants such as excess phosphates.

Many of the papers are concerned with the disposal of solid and liquid wastes. The Environmental Protection Bill proposes, in Part Two, a structure of waste regulation authorities based on local control of the licences issued to contractors, subject to the power of the Secretary of State to give directions on the terms and conditions of any licence. The design of waste disposal sites will require basic geological and hydrogeological information, and the use of barriers to prevent leachate from reaching underground water supplies.

Waste disposal has, therefore, to be reviewed in a very wide context. There is, certainly, now a need for an almost geographical insight if we are to advance from the original concepts which operated when the systems of sewage disposal in the U.K., combined systems to carry rainwater as well as sewage, were first installed. The calculation was that heavy rainfall could be accommodated by holding capacity in sewage treatment works, and that increased river flow would keep pollution at a low level. These concepts have to be refined. Geographical factors certainly complicate harmonization of standards in the European Community and the difference between a small island with a temperate climate, and a continental land mass with central mountain ranges. The latter have big catchment areas that accumulate snow in winter and release water in the summer by melting into massive rivers. As a result these have less extreme fluctuations and in the most difficult times flows up to two orders of magnitude greater than rivers such as the Thames, so pollutant dispersal is easier. But when air quality is being considered the situation is reversed. In continental areas air movements are much more restricted, and inversion conditions are more likely, so there is need for stricter controls on gaseous emissions.

However, it is easy to let such considerations bias approaches to waste disposal too strongly. On the one hand the waste removal properties of the Rhine and Danube have now been grossly over-exploited, and on the other, the U.K. has been taken to task for the sulphur it has for a long time disposed of into the air which so quickly takes it away but which deposits it elsewhere.

Waste disposal involves a great number of considerations. If these are not properly appreciated there is likely to be trouble, either immediately, or at some time in the future, as we have seen in the notorious Love Canal episode in the U.S.A. and in the current concern over the methane gas generated by past landfill operations in the U.K. The only solution is the use of powerful, scientifically based technology by which the best options can be developed. The papers in this book may help by identifying the work that is needed to provide data for decision between alternatives.

There remains another option: a changing lifestyle and reducing consumption and waste. It brings to the forefront the contribution which should be made by economists and social scientists to environmental protection. Otherwise scientists and engineers will be left with an ever increasing amount of cleaning up.

REFERENCES

Andrews, M.J. (1984) Thames estuary: pollution and recovery. In *Effects of pollutants at the ecosystem level* (ed. P.J. Sheehan, G.C. Butler and P. Bourdeau), pp. 195–227. Chichester: Wiley.

Hardin, G. (1977) The tragedy of the commons. In *Managing the commons* (ed. G. Hardin and J. Baden), pp. 16–30. Freeman: San Francisco.

Ministry of Agriculture and Fisheries Annual Report (1986–87) *Disposal of waste at sea*, p. 57. London: H.M.S.O.

Owens, N.J.P., Cook, D., Colebrook, M., Hunt, A. and Reid, P.C. (1989) *J. mar. Biol. Ass. U.K.* **69**, 813–821.

Pochin, E. (1977) Problems involved in developing an idea of harm. ICRP Publication No. 27. *Ann. ICRP* **1**, 1–24.

Ratcliffe, D. (1970) Changes attributable to pesticides in egg breakage frequency and eggshell thicknesses in some British birds. *J. appl. Ecol.* **7**, 67–115.

Royal Commission of Environmental Pollution (1975) 5th Report. *Air pollution control: an integrated approach*. London: H.M.S.O.

Warner, F. (ed) (1983) *The assessment and perception of risk*, pp. 158–194. London: The Royal Society.

1

Pollution and ecosystems

A.D. Bradshaw

1.1 INTRODUCTION

Wastes are an inevitable part of human activity. They are either a by-product of initial production processes, or they arise when objects or materials are discarded after they have been used. In a similar way, wastes are also an inevitable part of the natural world. Growth is a production system for living materials, which produces wastes as a by-product. What is more significant is that all living organisms, when they die, ultimately become waste. Despite this, nature appears to be able to deal with its own wastes very effectively.

Of course, the natural world is also the recipient of human waste materials. Many of these are completely different from natural materials. It would appear that most environmental problems arise because the natural world cannot cope with these novel substances.

The natural world is therefore a crucial starting point for any understanding of the environmental problems that waste materials can cause, and the ways in which they can be treated. The natural world is, however, too large and complex to be considered as a whole. The unit that has to be considered is the ecosystem, the combination of plants, animals, soil and climate that occur and interact together in a selected place.

The first essential principle of an ecosystem is that its components interact together in many complex ways. The outlines are shown in Fig. 1.1. The situation is very complex, as in each part of any ecosystem there are innumerable species or equivalent components, which themselves interact. The stability that appears in nature is therefore not the outcome of a static situation, but of a dynamic equilibrium. As a result, radical changes in an ecosystem can occur even if only one small component is disturbed.

The Treatment and Handling of Wastes
Edited by A.D. Bradshaw, Sir Richard Southwood and Sir Frederick Warner
Published in 1992 by Chapman & Hall, London, for The Royal Society
UK ISBN 0 412 39390 5, USA ISBN 0 442 31461 2

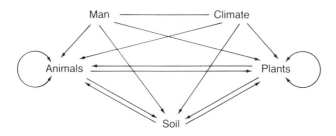

Figure 1.1 The components of an ecosystem and the interactions between them.

The second essential principle is that the components of an ecosystem are linked together by a flow and cycling of materials. Materials and energy cycle between one component and another. Without this transfer, growth in the ecosystem would be impossible. The relations may be quite complex even when only a single element is considered and species differences are omitted (Fig. 1.2).

The third principle, which of course applies to all matter, but which was first pointed out with great advocacy by Commoner (1971) to be particularly important when wastes are being disposed of, is that *'everything must go somewhere'*. It is very easy to get rid of a waste into the environment with the belief that it will just disappear. This it may apparently do, but the material or its components must actually move through ecosystems and end up in definable situations, where problems may arise.

1.2 WASTES FROM NATURAL SYSTEMS

The natural world is a very interesting model for any consideration of waste disposal. Large quantities of wastes are produced by natural ecosystems. However, in every ecosystem decomposition and cycling processes are well established. Although there are effectively no natural toxic wastes, without decomposition and cycling there would be chaos. First, there would be immense accumulation of wastes, which would then lead to a slowing down and ultimate cessation of growth within the ecosystem, because the production of ecosystems, which may be substantial, is dependent on the recycling of components, especially mineral nutrients, brough about by waste decomposition.

The major site for this is in soil, which contains a vast range of organisms carrying out decomposition processes. In a square metre of normal soil there are about 1000 species of microscopic animals and 1000 species of micro-organisms involved in decomposition. In a single cubic centimetre of soil there are at least 1000×10^6 bacteria and 60 m of fungal mycelia. The diversity of species and of the materials they decompose is immense (Swift *et al.*, 1979).

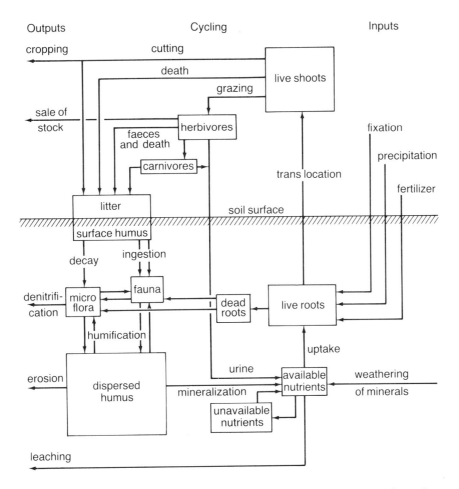

Figure 1.2 A diagrammatic representation of the flow of nitrogen through a terrestrial ecosystem. The size of the compartments indicates approximately the size of the store of nitrogen they contain.

A good example of what occurs within an ecosystem is provided by a temperate forest (Likens *et al.*, 1977). These produce large amounts of waste (over 10 000 kg ha^{-1}†) annually, particularly in the form of dead leaves, twigs and branches, which fall to the ground as litter. By decomposition, major supplies of nutrient are released from the litter and are taken up immediately by the forest (approximately 100 kg N, 50 kg Ca, 50 kg K, 10 kg P ha^{-1}). Without this, forest growth would be impossible, although the cycle

† 1 hectare = 10^4 m^2.

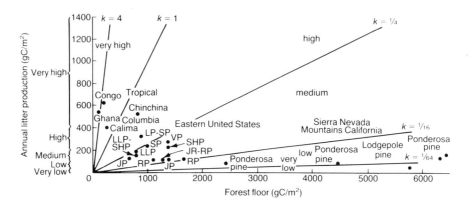

Figure 1.3 A comparison of the annual litter production and the amount of litter remaining on the floor of a range of different forests. The lines correspond to different rates of decomposition of the litter (from Olsen, 1963).

is topped up by contributions from rock weathering and the atmosphere. At the same time 4500 kg of C is released as carbon dioxide, which joins the general atmospheric pool used for further plant growth.

Decomposition, as well as production, is affected by the environment. So in inhospitable environments such as cold, wet, northern forests, there is not only slow growth and therefore low production of waste, that is, litter, but also very slow decomposition. In these conditions large amounts of un-decomposed litter may accumulate on the forest floor. In tropical forests the situation is reversed. In all cases an equilibrium is reached, related to the rate of decomposition (Olsen, 1963) (see below and also Fig. 1.3). In forests in the same climatic region, productivity can be directly related to rate of organic matter breakdown and nutrient release (Nadelhoffer *et al.*, 1983).

Some plant materials break down very quickly, particularly those with a low C:N ratio. At the same time, there are several recalcitrant fractions, particularly lignin. These may take very many years to decompose and are the major constituents of humus in soils. Nevertheless, they ultimately also break down and release the plant nutrients they contain.

The decomposition processes in nature are very complex (Swift *et al.*, 1979). Nevertheless, they are very positive and can be modelled (McGill *et al.*, 1981) to understand their exact importance. The essential conclusion, however, that there are no waste products in nature that are not decomposed and recycled, is self evident. There are complex sets of biological systems, which ensure that all natural wastes are decomposed so that few or no adverse effects of wastes can be found. As a result, maximum benefit for living organisms is derived.

1.3 WASTES FROM ARTIFICIAL SYSTEMS

If the wastes produced by human beings had no negative properties, they would not be a problem, but they have both physical and chemical properties. It is the interaction of these with the natural and artificial ecosystems, which occupy the surface of the Earth, that defines the problem set by any waste, both in nature and magnitude.

1.3.1 High bulk, solid wastes

Before examining those wastes that are the major cause of our problems, there is one particular type of waste which requires separate consideration. This is the high bulk, solid, inert wastes mainly produced by mining, which are normally placed on land. These give rise to the special problems of derelict land. If they are effectively inert, they have few specific environmental effects, except that they extinguish existing ecosystems by their physical presence. Such effects cannot be dismissed, but they are usually self-contained, and at the same time the problems can be treated. New ecosystems can be created or the previous ecosystems restored.

The problems of derelict land and the techniques for its reclamation have been discussed recently (Bradshaw, 1984). The key to the process is to charge up the newly developing ecosystem, which starts in a very skeletal condition, with enough nutrients, especially nitrogen, to ensure that sufficient is recycled each year to sufficient for current growth (Bradshaw, 1983). Interestingly, one major way to achieve this is to provide a heavy application of a waste, sewage sludge (Byrom and Bradshaw, 1991; Hall, Chapter 5).

1.3.2 Toxic wastes

Inert wastes do not present many difficulties. The major problems that lie in the great range of other wastes are: (1) toxic or otherwise troublesome; (2) may spread into other ecosystems; (3) may cause pollution, which can be best defined as the release of a substance into the environment, at least in part as the result of man's activities, which has a deleterious effect on living organisms.

These wastes may be naturally occurring materials, eg. CO_2, NO_3, even simple organic matter, but released into the environment at higher than normal levels. Usually, however, they are substances not normally found in nature. They will have some sort of physical presence, but it is their chemical presence which matters. This involves both the immediate characteristics of

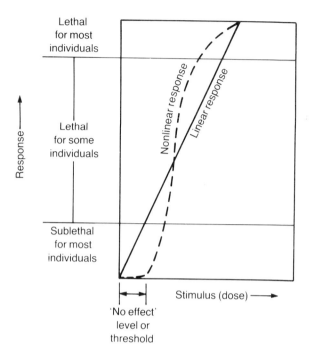

Figure 1.4 A generalized model of the response of a species to a toxic material (from Sheehan, 1984).

the wastes as they are released, and the derived characteristics developed after release and passage through ecosystems.

1.3.3 Direct effects

The immediate effects of a waste are clearly very important. They can only be assessed by careful experimentation in relation to the many organisms of the living world. This constitutes the large and well developed science of ecotoxicology. The crucial point is that different materials have different effects on different targets, which are the many different species with which the waste can come in contact.

Each species will have its own concentration/response curve in relation to an individual substance (Sheehan, 1984) (Fig. 1.4). To determine responses sufficiently precisely so that prediction is possible, detailed experimentation is necessary, in which all attributes are checked under different conditions. In Fig. 1.4, for example, the character being assessed is survival. Yet in many situations growth may be more important. This is very clear in the recent experience of assessing the effects of SO_2. Originally, the important

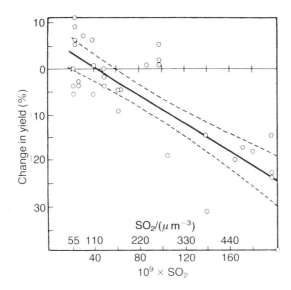

Figure 1.5 The response, measured in terms of yield, of ryegrass (*Lalium perenne*) to different concentrations of SO_2. Visible injury only appears when SO_2 levels are above 500 μg m^3 (from Roberts, 1984).

effects of SO_2 on plants were considered to be only those of visible injury, which occurred at levels above about 500 μg m^{-3}. However, all recent work has shown serious invisible effects on growth at much lower levels (Fig. 1.5). However, this is not all; the effects of SO_2 on growth are considerably influenced by season, being at least doubled in early spring and reduced to nothing in summer (for detailed discussion see Roberts (1984)).

The environment into which a waste is released can also have major mitigating or enhancing effects. When SO_2 is released into the atmosphere it has major effects as acid rain. This acidity can obviously be neutralized by alkalinity in the ground or in the water. What is more subtle is that its effects on living organisms may be mitigated by calcium without neutralization (Howells, 1984) (see Fig. 1.6). These few examples show how difficult is the critical assessment of the toxicity of a material. The complexity of biological systems limits the possibility of prediction (for further discussion see Moriarty (1983)).

1.3.4 Derived effects

After release, wastes: (1) can be acted on by environment; (2) can move into and through ecosystems as a whole or as components, which gives rise to many further complications.

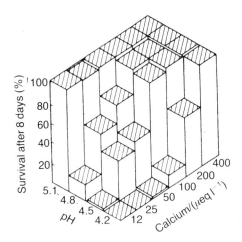

Figure 1.6 The interaction between calcium and pH in their effects on the survival of fry of brown trout (from Howells, 1984).

(a) Transformations

The first possibility is that transformations of the material can occur. These are related to the chemistry of the material and the action of environment. Many of these can result in the material being changed to a less toxic form. When chromate wastes are released into aquatic environments, they are commonly reduced to the chromic form, which is much more readily bound by organic and other materials and rendered unavailable (Breeze, 1972). The oxidation of SO_2 to SO_3 and therefore to sulphate, eliminates its toxicity, and was an important argument for the dispersal of sulphur emissions by tall stacks. However, as we now know, this policy failed to take account of the acidifying effects of the sulphate.

It is very possible for transformations to occur which increase the toxicity of a material. The most spectacular example of this is the transformation of mercuric compounds, which are not normally poisonous, to methyl mercury, which is extremely poisonous and readily taken up by living organisms (World Health Organization, 1989). This can occur in aquatic systems by biological or by chemical means. The appalling misery and deaths that occurred in the human population consuming fish from Minimata Bay in Japan is a sorry tale for waste disposal, and an object lesson that the effects of transformations must be properly understood.

(b) Interactions

The second possibility is that once a material is released, its major effects are due not to their primary effects, but to interactions occurring within

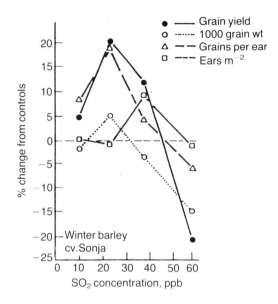

Figure 1.7 The positive effect on SO_2 fumigation at low concentrations on the yield of grain and other characters of winter barley growing under field conditions (from McLeod *et al.*, 1988).

the ecosystem. What actually happens relates to the specific characteristics of the ecosystem affected, and are therefore difficult to predict. Three examples will show this.

The invisible injury caused by SO_2 has already been considered (Fig. 1.3). When, however, the effects of SO_2 on cereal crops are examined under field conditions, increases in yield can be found at low concentrations of SO_2 (McLeod *et al.* 1988) (Fig. 1.7). In this experiment the SO_2 had clearly inhibited the growth of fungal pathogens, particularly brown rust, *Puccinia hordei*. This is in keeping with the known effects of SO_2 on other fungal pathogens, such as black spot of roses. At higher concentrations, it must be noted, the expected adverse effects of SO_2 reappeared.

When the effluents from sewage treatment plants are discharged into rivers it is usually considered that, being fully treated, they will not have any adverse effects. Yet although the organic material has been removed, the nutrients, particularly nitrogen and phosphorus, remain. Although these are naturally occurring, they can cause increased growth of algae. In a specific water system, such as the Norfolk Broads, the effects can be much more complicated because of the interactions with other impacting factors as well as with different components of the ecosystem. The result can be an almost complete breakdown of the ecosystem (George, 1977) (Fig. 1.8).

The toxicity of heavy metals is well appreciated. High levels of metal

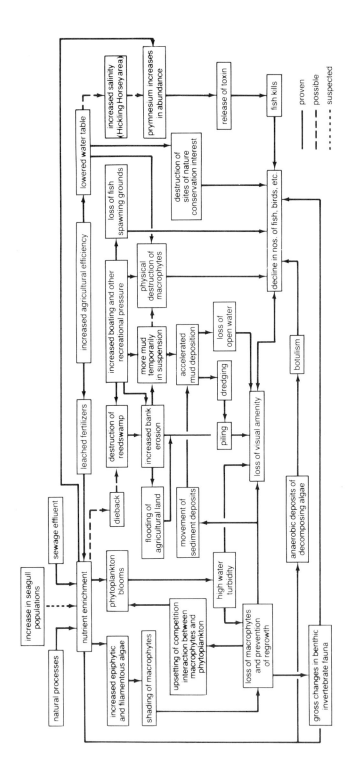

Figure 1.8 The complex interactions involved in the effects on the ecosystems of the Norfolk Broads caused by the discharge of treated sewage effluent containing high levels of nutrients (from George, 1977).

Table 1.1 Litter acculumlation in relation to metal concentration in woodlands at different distances from a metal smelter (from Hutton, 1984).

Wood	Litter mass $(g\,m^{-2})$	Distance from smelter/km	Metal content of litter ($\mu g\,g^{-1}$ dry mass) Cadmium	Lead	Zinc
Moorgrove	8345 ± 2131	2.5	23	1052	764
Blaise	13160 ± 1411	2.9	32	721	1844
Hallen	8343 ± 598	2.9	62	2179	1469
Haw	7910 ± 640	3.1	98	1545	2814
Leigh	1784 ± 157	6.8	7.2	191	169
Wetmoor	913 ± 100	23.0	1.5	44	80
Midger	3104 ± 268	28.5	5.7	103	202

contamination originating from the disposal of metalliferous mine wastes on land can lead to areas completely devoid of vegetation. But there are other areas, such as those affected by the aerial fallout of dust from refineries, where metal contamination is less high. In these areas plants may not be killed, but there is distinctly reduced growth. Whereas several factors may be involved, one most conspicuous feature is the accumulation of organic matter (Hutton, 1984) (Table 1.1). Metal pollution can be shown to prevent organic matter breakdown almost completely. Since this in turn will prevent the normal cycling of nutrients, it is suggested to be a major cause of poor growth in such polluted situations (Tyler, 1972).

(c) Flows and accumulation

As there are major flows of materials through ecosystems, polluting materials can also move through ecosystems and perhaps accumulate in certain places. What may happen is best understood by a simple input/output model (Fig. 1.9). The material in a given compartment or part of the ecosystem is related to the rate of input and the rate of output. The latter, which can be due to outflow, dilution, degradation, neutralization or complexing, can be quantified by a rate constant for loss, the reciprocal of which is the residence time. At equilibrium the critical factor is the rate constant for loss.

Complex models can be built up linking different parts of an ecosystem, from which the detailed movement and fate of pollutants can be followed. An analysis of the movement of the different heavy metals within the soils described in Table 1.1 shows that zince and cadmium are being lost from these soils whereas lead is being retained (Martin and Coughtrey, 1987). Similar models are very important in following radionuclide movement and transfer (Jackson and Smith, 1987).

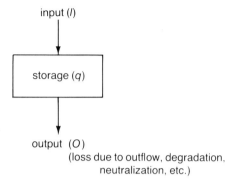

input (*I*)

storage (*q*)

output (*O*)
(loss due to outflow, degradation, neutralization, etc.)

If λ = rate constant for loss (with units of t^{-1})

$$\frac{dq}{dt} = I - \lambda q$$

At equilibrium $q = \frac{I}{\lambda}$

Figure 1.9 A simple input–output model. At equilibrium the amount in storage is determined simply by the input/rate constant for loss.

(d) Matrix of effects

There are now many other such refined analyses. But the essential simplicities of the situation must not be forgotten. For a given input there is a matrix of interactions, related to the openness of the ecosystem and the stability of the polluting waste material (Fig. 1.10). This may seem an oversimplification of what may happen, yet the extremes draw attention to where the essential problems of waste disposal lie. On the one hand, where a degradeable pollutant is discharged in an open ecosystem, as when organic matter is discharged into a stream, although there may be a local effect on critical parameters such as biological oxygen demand (BOD), this disappears rapidly

		Type of pollutant	
		Degradeable	Stable
Type of ecosystem	Open	Small	Medium
	Closed	Medium	Large cumulative

Figure 1.10 The different outcomes arising from the matrix of possible interactions when different types of waste material are discharged into different types of ecosystems.

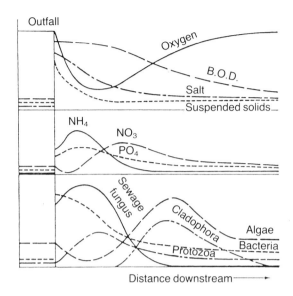

Figure 1.11 A diagrammatic representation of the effects of an organic discharge into a river and the changes occurring as the material passes downstream. There is a rapid recovery in oxygen levels, but this is accompanied by an excessive growth of algae (from Hynes, 1960).

downstream (Hynes, 1960) (see Fig. 1.11). On the other hand, where a stable pollutant is discharged into a closed ecosystem, as when metalliferous waste is discharged onto land, there are major damaging effects that may well remain for 200 years or more. This can be readily seen in the lead mining areas of mid-Wales, many of which are 300 years old, or in the great copper mine at Rio Tinto, begun over 2000 years ago by the Phoenicians.

In the latter examples, if the waste or its components were completely immobile and the receiving ecosystem could therefore be regarded as being completely closed, the problem would at least be limited to the receiving area. Unfortunately, heavy metals in mining wastes have some mobility. As a result metals can leak out of mining wastes by leaching for very long periods of time, almost indefinitely. It is clear that the disposal of toxic non-degradeable wastes is fraught with difficulties.

(e) Different components

Most waste materials are complex and contain many different components. This can be a source of different problems. The various components may have different physical behaviours. Lead and zinc commonly occur together in the same mine waste, yet zinc is released from such wastes much more

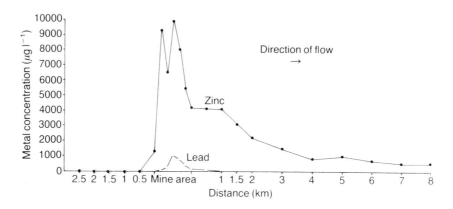

Figure 1.12 The zinc and lead found in the sediments of a stream passing through the wastes of a disused lead–zinc mine containing approximately equal quantities of lead and zinc. The zinc is much more mobile than the lead.

readily than lead (Fig. 1.12). As a result, in regions of old lead mine workings, such as in mid-Wales, present-day river pollution is due to zinc rather than lead.

When organic matter is discharged into a river, it used to be considered that the river had recovered when BOD had returned to normal, signifying that the organic matter had decomposed. Organic matter contains plant nutrients such as nitrogen and phosphorus, which are released into the water together with carbon dioxide, and they cause an increased growth of algae and higher plants (Fig. 1.11), which may actually choke the water body.

Sewage sludge is essentially a natural material. As such, it should be possible to be disposed of very easily, and used to return fertility in the form of lost nutrients to agricultural land. Its use, however, is limited by its content of heavy metals emanating from industrial and other sources (Hall, Chapter 5). It is unfortunate that we allow the mixing of such different materials together in the collection process, and so preclude the effective disposal of a valuable waste.

(f) Bio-accumulation

Ecosystems consist of food chains, or food webs, in which material passes from the lower trophic levels, starting with plants, to the higher trophic levels, firstly herbivores and then carnivores. As a result, particular pollutants can become concentrated at higher trophic levels, if they are stable and are not excreted by the organisms concerned. What happens in practice can be the outcome of simple concentration processes related to food consumption. This has had considerable consequences for wildlife. The

most notable example is the accumulation of chlorinated hydrocarbons in
birds of prey, which lead to eggshell thinning and serious population declines
in the 1940s (Moriarty, 1983).

   Bio-accumulation has, however, many complexities in practice. Obviously, it is determined by the balance between intake and retention, the former
related to food and the latter to the interaction between the substance concerned and the physiology of the species. In an area of grassland contaminated
by copper and cadmium from an adjacent copper refinery, metal levels were
examined in voles (herbivores) and shrews (insectivores) and related to their
respective diets (Hunter *et al.*, 1987) (Fig. 1.13). Despite very high levels in
the vegetation there is no sign of accumulation of copper in either species,
because copper is not retained and is well regulated in most animals by
excretion. Cadmium, by contrast, shows marked accumulation, which is
especially noticeable in the insectivorous shrews because of their diet, sufficient to cause obvious kidney damage. By using the herbivorous voles to

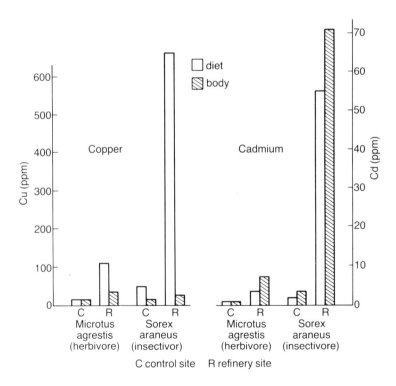

Figure 1.13 The concentrations of copper and cadmium in the diet and in the bodies
of two contrasting species of small mammal (herbivore and insectivore). There is
substantial accumulation of cadmium but not copper, particularly in the insectivore
(from Hunter *et al.*, 1987).

provide the comparison, cadmium has a food chain transfer potential 145 times that of copper.

Retention of a particular pollutant can, however, be very different in different species. The 'half-life' for methyl mercury can vary from 8 days in the mouse, 70 days in man, to over 1000 days in some fish (Clarkson, 1972). Although bio-accumulation is not easy to predict, it must be considered as an important possibility whenever wastes are being disposed of.

(g) Physico-chemical accumulation

Physico-chemical accumulation will depend on many different subtleties of the environment into which a waste is discharged, and there is, therefore, no need to examine the processes in detail. However, the possibility of quite startling local accumulations of particular pollutants, such as in river and estuary sediments, arising from waste discharge must always be considered, with their potentially serious consequences. Major problems of disposal are continuing to arise, for instance, with river dredgings in both the U.S.A. and Europe. At the same time this accumulation can be very patchy (Little *et al.*, 1987).

1.4 CONCLUSIONS

It is only possible to understand the environmental effects of a waste if its effects are examined and understood in the context of the ecosystems which it may affect. The natural environment is exceedingly complex, and the assumption of an ecosystem approach does little to simplify it. It does, however, make more clear the possible behaviour of the material and its components, its effects on the assembly of living organisms with which it comes in contact, and the range of problems which may be encountered as a result.

If a material is degradable to give harmless products, its discharge into the environment is unlikely to cause problems, although there may be local effects. If the material is not degradable, the situation is much more serious, because everything has to go somewhere. Because different small inputs can combine to give major problems, waste disposal can no longer be thought of as a local problem. A first step must therefore be, at least with the more toxic materials, to carry out careful audits to discover what are the major inputs into the environment. This is being done for major wastes such as carbon dioxide, for obvious urgent reasons. But it has also to be for other wastes with potentially serious effects, such as mercury and cadmium. The results can reveal unexpected sources, including, as for cadmium, a whole range of inadvertent ones (Hutton and Symon, 1987).

If the material is natural it is easy to think that it will be swallowed up by nature. This may occur, but there is the possibility that in the process of

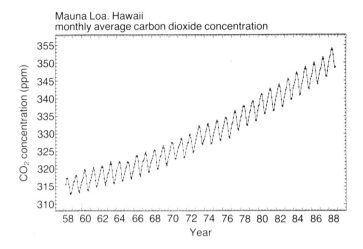

Figure 1.14 The progressive accumulation of carbon dioxide in the atmosphere over the last thirty years shown in the monthly averages at Mauna Loa, Hawaii. Every summer there is a reduction in concentration but this is more than cancelled out every winter.

dispersal it or its components will, as with N and P from organic discharges into aquatic environments, cause extra growth and adverse effects on particular ecosystems. But there is also the possibility that nature has only a limited capacity to absorb the material being discharged, and that the limits are reached. This is very clear now from the records of carbon dioxide levels in the atmosphere (Fig. 1.14). The annual oscillations show the capacity of the ecosystems of the Norther Hemisphere in the summer to absorb the CO_2 being produced. But in the winter this capacity is greatly exceeded, leading to the progressive rise in CO_2 which gives such cause for concern (Mason, 1989). Nitrogen in the form of nitrate is another example. It is essential, therefore, that when natural materials are discharged into the environment, the amounts and concentrations of the material and its derivatives, and the reactions of affected ecosystems, are carefully monitored.

If the material is unnatural the problems are much greater. It is essential to know the characteristics of the material precisely. The first criterion is its toxicity, exactly what effects can it have? To find this out may take a great deal of careful work. The second criterion is its mobility, how will it move through ecosystems into which it enters? Where will it end up? Experience has shown that this may be both difficult to predict and difficult to quantify. Complex pathways may be involved, so that effects may appear at a considerable distance, in geographical and biological terms, from the source.

The damage that may be caused by a waste can only be understood by work that takes account of its behaviour in whole ecosystems. This may involve complex and tedious work. But we would be acting irresponsibly, as everything has to go somewhere, if we did not know where that somewhere was.

REFERENCES

Bradshaw, A.D. (1983) The reconstruction of ecosystems. *J. appl. Ecol.* **20**, 1–17.

Bradshaw, A.D (1984) Land restoration: now and in the future. *Proc. R. Soc. Lond.* B **223**, 1–23.

Breeze, V. (1973) Land reclamation and river pollution problems in the Croal valley caused by waste from chromate manufacture. *J. appl. Ecol.* **10**, 513–525.

Byrom, K. and Bradshaw, A.D. (1991) The potential value of sewage in land reclamation. In *Alternative uses for sewage sludge* (ed. J.E. Hall), pp. 1–20. Medmenham: Water Research Centre.

Clarkson, T.W. (1972) Recent advances in the toxicology of mercury with emphasis on the alkylmercurials. *Crit. Rev. Toxicol.* **1**, 203–234.

Commoner, B. (1971) The closing circle. New York: Knopff.

George, M. (1977) The problem of the Broads. *Trans. Norfolk Norwich Nat. Soc.* **24**, 41–53.

Howells, G.D. (1984) Fishery decline: mechanisms and predictions. *Phil. Trans. R. Soc. Lond.* B **305**, 529–547.

Hunter, B.A., Johnson, M.S. and Thompson, D.J. (1987) Ecotoxicology of copper and cadmium in a contaminated grassland ecosystem III Small mammals. *J. appl. Ecology* **24**, 601–614.

Hutton, M. (1984) Impact of airborne metal contamination on a deciduous woodland system. In *Effects of pollutants at the ecosystem level* (ed. P.J. Sheehan, D.R. Miller, G.C. Butler and P. Bourdeau), pp. 365–375. Chichester: Wiley.

Hutton, M. and Symon, C. (1987) Sources of cadmium discharge to the UK environment. In *Pollutant transport and fate in ecosystems* (ed. P.J. Coughtrey, M.H. Martin and M.H. Unsworth), pp. 223–238. Oxford: Blackwell.

Hynes, H.B.N. (1960) *The biology of polluted waters* University of Liverpool.

Jackson, D. and Smith, A.D. (1987) Generalised models for the transfer and distribution of stable elements and their radionuclides in agricultural systems. In *Pollutant transport and fate in ecosystems* (ed. P.J. Coughtrey, M.H. Martin and M.H. Unsworth), pp. 385–402. Oxford: Blackwell.

Likens, G.E., Bormann, F.H., Pierce, R.S., Eaton, J.S. and Johnson, N.M. (1977) *Biogeochemistry of a forested ecosystem*. New York: Springer-Verlag.

Little, D.I., Howells, S.E., Abbiss, T.P. and Rostron, D. (1987) Some factors affecting the fate of estuarine sediment hydrocarbons and trace metals in Milford Haven 1978–82. In *Pollutant transport and fate in ecosystems* (ed. P.J. Coughtrey, M.H. Martin and M.H. Unsworth), pp. 55–88. Oxford: Blackwell.

McGill, W.B., Hunt, H.W., Woodmansee, R.G. and Reuss, J.O. (1981) Phoenix – a model of the dynamics of carbon and nitrogen in grassland soils. In *Terrestrial nitrogen cycles* (ed. F.E. Clark and T. Rosswall), pp. 49–116. Stockholm: Ecological Bulletins.

McLeod, A.R., Roberts, T.M., Alexander, K. and Cribb, D.M. (1988) Effects of open-air fumigation with sulphur dioxide on the growth and yield of winter barley. *New Phytol.* **109**, 67–78.

Martin, M.H. and Coughtrey, P.J. (1987) Cycling and fate of heavy metals in a contaminated woodland ecosystem. In *Pollutant transport and fate in ecosystems* (ed. P.J. Coughtrey, M.H. Martin and M.H. Unsworth), pp. 319–336. Oxford: Blackwell.

Mason, J. (1989) *The greenhouse effect: the scientific basis for policy*. London: The Royal Society.

Moriarty, F. (1983) *Ecotoxicology*. London: Academic Press.

Nadelhoffer, K.J., Aber, J.D. and Melilo, J.M. (1983) Leaf-litter production and soil organic matter dynamics along a nitrogen-availability gradient in Southern Wisconsin (U.S.A.). *Can. J. Forest. Res.* **13**, 12–21.

Olsen, J.S. (1963) Energy storage and the balance of producers and decomposers in ecological systems. *Ecology* **44**, 322–331.

Roberts, M.J. (1984) Long-term effects of sulphur dioxide on crops: an analysis of dose-response relations. *Phil. Trans. R. Soc. Lond.* B **305**, 299–316.

Sheehan, P.J. (1984) Effects on individuals and populations. In *Effects of pollutants at the ecosystem level* (ed. P.J. Sheehan, D.R. Miller, G.C. Butler and P. Bourdeau), pp. 23–50. Chichester: Wiley.

Swift, M.J., Heal, O.W. and Anderson, J.M. (1979) *Decomposition in terrestrial ecosystems* Oxford: Blackwell.

Tyler, G. (1972) Heavy metals pollute nature, may reduce productivity. *Ambio* **1**, 52–59.

World Health Organisation (1989) *Mercury – environmental aspects*. Environmental Health Criteria No. 86. Geneva: World Health Organization.

2

The nature of the waste problem: a question of prevention

M. Backman and T. Lindhqvist

2.1 INTRODUCTION

In principle a similarity can be seen between the strategies that have been used to attack waste problems. In most cases wastes are considered independently of their sources of origin, their extent and their content. Thus the attack on the problems has been concentrated and shifted in both space and time to the product, while the causes of the onset of the problems have seldom been analysed or remedied.

Obviously, the above strategies give no basis for a sustainable development. The fundamental mismanagement of the resources of the earth, reflected in wastes, cannot be corrected by a re-active strategy that mainly moves the problems around. At the same time, however, it is the mismanagement mentioned that provides the key to a more pro-active strategy. As raw materials correspond to a certain value, and because the costs of treating wastes are increasing, there is an obvious economic incentive to remedy the origin of the problems. By amplifying this incentive in various ways, for example, with economic steering instruments, a waste prevention strategy can be made to substitute the earlier end-of-pipe strategies.

The Treatment and Handling of Wastes
Edited by A.D. Bradshaw, Sir Richard Southwood and Sir Frederick Warner
Published in 1992 by Chapman & Hall, London, for The Royal Society
UK ISBN 0 412 39390 5, USA ISBN 0 442 31461 2

2.2 HISTORICAL DEVELOPMENT

2.2.1 Background

A historical discourse on which discussion about waste can be based, could, of course, be made very extensive. However, a brief review of only a few decades, can provide an adequate point of departure for an understanding of the present problems and of future recommendations and discussions.

During the post-World War II period of growth, environmental issues were not regarded as an obvious limiting factor. Waste products from industrial activities and from consumerism were allowed to contaminate the air, water and the soil without much control. Legislation and other regulations were aimed mainly at the organization of household waste collection, and at the improvement of the hygienic aspects in conjunction therewith.

In step with the expansion of industrialization and the improvement of welfare, an understanding of the negative aspects of growth also increased. The enormous increase in the amount of pollutants and wastes made the threat to our common natural resources ever more obvious. Initially, this threat was met with the dilution strategy with higher smoke stacks and longer drains as significant signs. Consequently, water and air pollutants were spread over large areas, also from dumps lacking adequate control and where open incineration was commonly used.

During the 1960s, filter technology, applied to smoke stacks and drains, came to be increasingly used. The pollutants could thus be concentrated into filter deposits, sludge and other concentrated forms. The growing need for locations for landfills began to decrease because more and more communities increasingly relied upon controlled incineration technology.

2.2.2 The epoch of centralized recycling

The 1970s saw a general growth in an understanding of the fact that the earth's resources are limited. The limits of growth began to be discussed, as also the strategies that had been in force till then and their inadequacy in safeguarding the long-term supply of environmental and natural resources.

Investments in centralized recycling technologies came to be regarded as the strategy that would be able to compensate the above short-comings. At that time there seemed to be no limit to what could be recycled and re-used with the help of advanced technology for sorting, separation and reworking; solid and hazardous wastes, sludges, filter deposits, etc.

It is obvious from the fundamental attitude, as formulated in 1975 by the Swedish government and parliament, how high the expectations for these advanced recycling techniques were; technology would make it possible to regard waste as *an asset for the community*. To take advantage of this expected asset, about 15 very costly, centralized plants were built for the breaking-up, separation, material extraction and re-use of wastes. The most advanced of these plants are now shut down as a consequence of unsatisfactory operation and very limited markets for the materials produced. Other plants are still operating, albeit as a relatively small fraction of their capacity.

Furthermore, it must be noted that the recovery achieved has been very poor. The quality of the ferrous metals that are produced is usually too low to be saleable to steel works. Where paper and plastic have been separated, these have also been found to be unsaleable due to high levels of contaminants. The compost produced can mainly be used only as the covering layer at landfills.

For the environment, the advanced recycling techniques are not beyond criticism. The compost produced often contains organic poisons and heavy metals, as well as scraps of glass and plastic. Moreover, the production process itself has been criticized because of the unpleasant odours that spread over the surrounding areas. The fuel fraction (RDF) contains approximately the same polluting substances as does the untreated waste, albeit in somewhat lower concentrations.

The greatest objection to the advanced recycling techniques is that the largest part of the treated waste must, nevertheless, be disposed of because of a lack of markets for the contaminated products produced.

As a result of the above mentioned experiences, no additional large investments have been made in centralized recycling technologies within Sweden since the beginning of the 1980s.

2.2.3 The epoch of incineration

For want of a functioning recycling technology, incineration was, for a considerable period during the 1980s, regarded as an attractive solution for the treatment of mixed waste. This technique, however, has recently become the focus of increasingly harsh criticism by researchers, environmental organizations and the general public, as well as coming under more stringent examination by the environmental authorities.

Apart from the discharge of dioxins and dibenzofuranes, the discharge from incineration plants of acidifying substances, polyaromatic hydrocarbons (PAH), chlorobenzenes, and many other chlorinated hydrocarbons, have been observed. The acidifying elements in the discharge contribute to the leaching of heavy metals and aluminium. Damage to vegetation in the vicinity of the plants has also been observed showing obvious gradients of damage to forest trees in the direction of the prevailing wind. The discharge of heavy metals is also greater in existing incineration plants than from plants that burn coal, oil, wood chips or peat.

By the use of advanced techniques for cleansing exhaust smoke, the airborne discharges can be reduced, but seen as a part of the total discharges (including the filter sediments, the condensates, the slag and ash), this cleansing process mainly redistributes the pollutants.

The incineration techniques do, however, also produce some positive effects on the environment. The volume of the waste, and thus the volume required for its deposition, is reduced considerably. This is obviously of great importance in regions with a high population density where it is almost impossible to find suitable places for a landfill. It must, of course, not be forgotten that places for deposition are still necessary, for the non-combustible material, slag and ash from the incineration plants.

Concerning the latter, it should be noted that combustion releases the metals from organo-metallic compounds. The leaching of cadmium, mercury, lead and zinc is usually greater from fine ash than from slag. On the other hand, other metals (e.g. copper, chromium, nickel and manganese) are leached to about the same extent or more, from slag than from fine ash. These facts make it essential to treat both the fine ash and the slag in such a way that the leaching or other unwanted distribution of the metals, is reduced to a minimum.

It the heat produced by an incineration plant is distributed via the community heating grid, the waste replaces another source of heat. If this alternative source is coal or oil, then the amount of discharge is less from the heat from waste incineration, both of sulphur and nitrogen oxides. On the other hand, and as previously noted, combustion is associated with a greater discharge of heavy metals, dioxins, polyaromatic hydrocarbons, chlorobenzenes, etc.

In recent years, waste-water heat, ground heat, ground-water heat, and in certain cases, solar heat have been utilized increasingly in community heating projects. On the community heating market, therefore, the sources of heat available are increasing while environmental demands are being made more stringent, resulting in a less favourable competitive status of the heat from waste incineration vis à vis, cleaner sources of heat.

The factors discussed above have resulted in termination of the building of new incineration plants in Sweden.

2.2.4 A future strategy

From the above discussion it becomes clear that over the years, considerable financial investment and great interest have been given to several technical possibilities for the final treatment of waste. All of these treatments, however, have involved important technical, environmental and/or economic limitations. The scale of these limitations is connected more with the quantity, the character and the composition of the waste, than to the choice of the method of treatment. The reason for this is that waste has been considered and treated as if it were independent or the sources of its production, its scale and its contents!

The strategy described obviously does not provide a basis for a long-term sustainable development. The fundamental faults in the way in which the earth's resources are utilized, as reflected in waste and pollution, cannot be corrected by technology, which in the best case can only shift the effects of the pollution burden in time and space. The expectations that it will be possible to correct, at a later stage, by law and stricter regulatory controls, all the environmental problems, which are the result of environmentally damaging raw materials, production methods, products and consumption patterns, are also entirely unrealistic.

The causes of the environmental disturbances due to waste treatment must therefore be sought, predominantly in earlier stages, i.e. in the product design and the production stages. The consumer also has a certain responsibility for the situation, by the way he handles and discards used products.

For these reasons, in Sweden, as in many other countries, there is a growing consensus concerning the need for planning for future work on the environment. The authorities involved, political parties, trade unions, researchers and environmental organizations are, in principle, in agreement that sustainable development must be based on a preventative strategy for the protection of the environment.

In the case of solid waste, this means that as far as possible the creation of waste must be kept to a minimum. By preventative measures, the waste, which is nevertheless created, should not contain environmentally damaging materials and should be prepared for re-use or recycling. Finally, it is essential that prerequisites are created for an effective sorting of re-usable materials at the source, and that measures are taken to ensure that a stable market exists for these materials.

In 2.3, the applications and potential of the preventative environmental strategy are illuminated both with respect to production and to wastes produced.

2.3 AN INCREASED RESPONSIBILITY

2.3.1 Introduction

A waste prevention strategy requires the broad participation of society, since, in addition to industry's activities, the professional and private activities of the citizens are of decisive importance. But obviously, a waste prevention strategy makes specific demands on the participation of industry in a process aimed at minimising the environmental impact both from the production itself and from the products that are used in society. The present aim is primarily for a participation extending beyond what is controlled by laws and decrees.

The role of the authorities, and current legislation, has so far mainly been concerned with the size and extent of certain environmental disturbances and, when they have already happened, the imposition of restrictions. However, legal procedures are slow and far from all environmental problems can be dealt with before major damage to our environment occurs. The generosity extended by the market economy to industry and to new products thus presumes an active, preventative responsibility on the part of industry in environmental questions. To develop and to refine steering instruments to activate such an environmental leadership should have a higher priority in a future preventative strategy for a sustainable development.

It should be emphasized that the steering instruments do not necessarily have to be initiated by the environmental authorities. Industrial leaders, who have become aware of purely economic disadvantages in the traditional, passive and defensive, environmental strategy, propose that companies establish their own internal environmental programmes to find economically favourable and environmentally sustainable solutions to their environmental problems, without awaiting directions from the authorities. However, only a few companies have succeeded in implementing systematic and purposeful programmes of a preventative nature.

The environmental responsibility of consumers must also increase in the future. Through their purchases, consumers exert a measure of control over the choice of products and thus they can contribute to the creation of markets for more environmentally sound alternatives. Also, through their general life style and through their handling of the products during and after the time of use, consumers can contribute to creating a more sustainable society. The consumer must not be led to believe that he is without influence in the field of environmental development.

2.3.2 Steering instruments

To create an increased protection of our common raw materials and environmental resources, different kinds of steering instruments can be used. All such instruments can be considered as having a common origin in the need to supplement the mechanisms of the market economy so that external environmental effects are taken into account to a greater degree when making industrial or private economy considerations. There are different ways of classifying these steering instruments, but one commonly used is shown in Table 2.1.

This report now concentrates on some of the steering instruments that can be expected to give steady impetus to a preventative strategy and to an increased industrial responsibility in the environmental sector. For increased clarity, those steering instruments that are believed to stimulate this responsibility for environmental pressure caused by industrial production are considered first. Afterwards, those steering instruments are described, which have similar functions where the products are concerned.

Table 2.1 Classification of steering instruments

Steering instrument	*Example*
Administrative	Prohibitions
	Regulations
Economic	Taxes
	Fees
	Economic support
	Deposit sytems
Informative	Propaganda
	Information
	Technical assistance
	Environmental product declaration
	Environmental labelling
	Waste exchanges
	Waste reduction audits
Agreements	Contracts with society
	Gentlemen's agreements

2.4 PRODUCTION

2.4.1 Economic instruments

A market economy requires that each activity carries its own total cost so that, for instance, communal environmental and natural resources can be shared in an optimal way. Immediate and future environmental effects, which are not assigned a real value in a business cost estimate, thus have to be brought into the economic considerations of producers and consumers by economic fees. The generally accepted Polluter Pays Principle (PPP) lends support to this type of steering instrument.

The application of the PPP is often built on the assumption that the real environmental costs can be assessed; a very complicated procedure in practice. Even if considerable means were spent on assessing the collective willingness to pay for environmental improvement to estimate the environmental costs, it is unlikely that such an assessment could be used to determine the size of an environmental fee. First, all decisions have to be subjected to political consideration of the total effects. Furthermore, there is the risk that the production is moved and the environmental problems 'exported' to countries with lower environmental ambitions. Another risk is that high fees may cause an increased concealment of actual emission conditions, which will also have a negative effect on the possibility of using other instruments to regulate the environmental pressures.

It follows that it should be practicable to subject many types of environmentally harmful raw materials and/or emissions to environmental fees, the fee in each individual case being kept low relative to prevailing purchasing and treatment costs. If this strategy is chosen there is small likelihood of the fee in itself having any obvious steering effect. However, this can be accomplished at a second stage if the collected fees are used for research and development as well as for economic support measures, with the collective aim of preventing the environmental pressure effects (environmental costs), which are covered by the original environmental fees.

Today's waste incineration plants and the landfills for solid and hazardous wastes may be taken as examples of activities that cause extensive environmental costs, which in accordance with the PPP should be charged to those who deliver their wastes to the different plants. However, if this principle is carried to the extreme, there is a clear risk of illegal dumping of the wastes, causing even greater pressure on the environment. By imposing an environmental fee that bears a reasonable relation to current costs of handling and treatment, one could collect sufficient funds for the initiation of active and purposeful programmes for research, education and the communication of information about new and better processes and technologies. The purpose would be to minimize the generation of wastes and thus the need for end-treatment plants.

2.4.2 Informative steering instruments

Many times it has been claimed that today's environmental problems, to a high degree, are caused by attitudes rather than by technical problems. If this is true, as many investigations indicate, then the importance of the information steering methods for solving the problems cannot be over-emphasized. It always takes a long time to break old traditions and to alter current attitudes and practices. If we are to succeed with waste prevention, then extensive information and educational efforts are needed to alter the directions of industrial activities and also the work of the authorities, various research and developing groups, the educational system etc.

An active pollution prevention programme is, at present, under development inside several industrial companies, mainly within the U.S.A. and northern Europe. However, it is usually the strong companies that are in a position to initiate and carry out their own programmes of research and development for reducing the wastes generated by their activities. Inside small- and medium-sized companies there is still a general lack of that strategic approach to environmental questions, which would contribute to the reduction of environmental problems without the intervention of the authorities. Considering the number of such industries and the resources available to the environmental authorities, this situation will assure continued environmental degradation if extensive resources are not spent on information and education and if technical assistance is not provided.

An informative instrument that should be developed and administered within companies and organizations, is the waste reduction audit. In line with a preventative strategy, the internal waste reduction audit functions as a tool for documentation of the types and quantities of the wastes produced and for identification and evaluation of various options for the minimization of the generation of wastes and pollutants. Regular waste reduction audits, thus become the internal steering instrument of the company directorate for localizing mismanagement of all resources of the company. They also ensure that waste minimization goals are attained.

Those companies that have instituted a waste prevention programme, including repeated waste reduction audits, have found economic advantages in a zone previously only associated with increased costs. By tackling environmental problems early in the production process, the following benefits, among others, can be gained.

1. Decreased costs of raw materials, energy and waste treatment.
2. Improved productivity and product quality.
3. Improved working conditions, reduction of sick leave, lowered personnel turn-over and improved recruiting possibilities.
4. Increased preparedness for future tightening of environmental law and emission controls.
5. An enhanced environmental image of value in public relations.

2.5 PRODUCTS

2.5.1 Background

In recent years, there has been an ever-growing awareness of the environmental impacts of products consumed by society. There are also strong indications that concern for products and their environmental impact will increase in the future. Although much attention has been devoted to the production processes and the reduction of the wastes from them, much less attention has been given to the products. At the same time, as various solutions for reducing the environmental impact of the production are implemented, there is the potential for a relative increase in the impact of the products.

The possibilities of influencing the environmental impact of a specific product are present during the whole life cycle of the product and thus the responsibility is shared among the various people handling the product. The producer, though, has unique possibilities of influencing the environmental qualities of a product already during the design phase. However, this assumes that environmental properties are considered when the product is designed and constructed and that these factors are assigned importance relative to other product properties that are striven for. Some of the goals that can be formulated for product development are the following.

1. To avoid at the different product phases (production, distribution, usage, end-treatment): (*a*) hazardous emissions; (*b*) hazardous elements.
2. To improve: (*a*) usage of raw material and energy; (*b*) the durability of the product; (*c*) the possibility of repairing the product; (*d*) the possibilities of recycling the product.
3. To facilitate: (*a*) the end-treatment of the product.

All these goals are not equally relevant for all kinds of products and can, in certain cases, be subordinate. For a product that consumes few resources and that has little environmental impact during production and final management, a long life need not be of overriding importance. Similarly, a total evaluation of the individual case must determine what is optimal for other characteristics of the product, such as the possibility of making repairs and its recyclability.

2.5.2 The environmental product declaration

Significant obstacles to a more environmentally oriented design are raised by the lack of awareness of the importance of the environmental impact of products as shown by constructors and industrial designers. There is also a need for more understanding of the effects that different construction decisions have on the environmental properties of the products. Consequently, considerable attention must be paid to research on the environmental impact of the products, and to the education of technicians and designers about these properties.

However, education and information will hardly be enough to guarantee a change towards sustainable development. Steering instruments of a more direct nature are also required to ensure that the environmental properties of products are assigned greater importance by the producers. The environmental product declaration (EPD) is a central concept in this context. The EPD is based on a principle that has been expressed somewhat differently by the authorities responsible for the environment. Basically, the producer, before starting the production of an article, should know which environmental effects are associated with that article and, in particular, how it is to be treated when it is expended. Linked to this principle, it should be reasonable to insist that this knowledge result in a consistent response. The knowledge about the environmental properties of the article must also be made clear to those who come in contact with the article during its life cycle, if they are to make rational decisions and to handle the product in an appropriate manner during its usage as well as when it is expended.

An EPD is a written description of the properties associated with a product which are basic from an environmental point of view. The format of an environmental product declaration must be adapted to each type of product, which can conveniently be done by cooperation between the concerned industry and the authorities. However, certain general features should be in common between most types of products. A complete EPD should contain the following:

1. a listing of all environmentally relevant substances contained in the product;
2. a description of the production steps that are important from an environmental point of view;
3. a description of the relevant, potential consumer exposure risks associated with proper and improper use of the product;
4. a description of the treatment of the product when it is expended with particular reference to the technical, economic and administrative prerequisites for reusing and recycling the product and its packaging. Furthermore, it should be stated how the product behaves during the normal methods of waste treatment such as landfilling, incineration and composting.

For more complex products, an EPD should also contain instructions for dismantling or scrapping, and a repair guide. An EPD functions as a steering instrument in several different ways. Its most important function is, during the product development phase, to make the various people involved conscious of the environmental properties of the article, thus using the producer's unique position to alter the construction of a product. By formalizing this procedure, the risk that the environmentally relevant properties are given a completely subordinate priority will diminish. Instead, the great creativity that is the hallmark of product development will be used for considering not only properties such as price, production speed, shape, colour, functionality, durability, potential market demands, etc., but also the properties that are important for the environmental impact of the product.

An EPD also gives important information for the authorities to use in planning. The possibilities of understanding and explaining causal relationships in the environmental field would be considerably improved by access to the information that is presented by a well-constructed system of environmental product declarations. Because of its relative complexity, the EPD is not particularly suited for direct use as a form of environmental product information to the great majority of consumers, but it is an excellent source of more consumer-friendly information.

2.5.3 System of environmental labelling

An EPD does not exclude other forms of environmentally oriented product information but can, in its various versions, be seen rather as a comprehensive concept, within which specifically fashioned environmental information can be contained. Different types of environmental labelling system can conveniently be looked at in this way.

During recent years, considerable attention has been paid to systems of

voluntary *positive environmental labelling* of the type that has existed for
more than ten years in the F.R.G. and which has been called 'The Blue
Angel'. The system gives information on individual products which fulfil
certain demands concerning environmental properties. The prerequisite for
a labelling system is that, within a specially defined range of products, there
exists a limited proportion of products with more favourable environmental
properties and that a shift of the consumption to products with these properties leads to environmental benefits.

Considerable efforts are needed to establish and sustain a well functioning, positive, environmental labelling system. They require international
collaboration so that the system can be established in a greater number of
countries. The significant exchange of goods between the countries also
motivates a coordination and harmonization of the environmental labels
because of consumers' interests as well as because of industrial interests.

It is important, for the functioning of a positive environmental labelling
system, that high standards for the awarding of the environmental label be
maintained and that the system is organized in such a way that it maintains a
high degree of credibility among the general public.

However, a voluntary, positive, environmental labelling system does not
fulfil all demands for environmentally related information that consumers
need. In particular, there is an appreciable need to know how the expended
products can best be treated; to what extent they are suitable for recycling,
if they can be composted, etc. A simple labelling in pictogram form would be
of great value in guiding the treatment of the expended products towards
methods with better environmental properties. The pictogram would state
what the product is made of and if the materials can be recycled; it would
also state whether the product can be composted, incinerated or landfilled,
and if it should be collected separately. It is, however, important that the
granting of a recycling symbol is linked to the demand for an efficient system
of collection and recycling as well as to a market for the recycled material.

2.5.4 Recycling

Recycling is an important way of improving the efficiency of usage of the
scanty resources of the world. It is, however, important that recycling is
not considered a goal in itself, but only a means of reaching a paramount
environmental goal. Neither should recycling lead to the neglect of waste
prevention. Because recycling procedures are never without loss, efforts
that, for instance, reduce a product's content of heavy metals, must not be
considered as alternatives to recycling operations. These aspects must be
considered as complementary.

Recycling requires an increased responsibility on the part of the producer.
Experience shows that the establishment of a successful recycling operation
requires that the producers take a decisive responsibility for the recycling

and ensure that there is a market for the recycled materials. In addition, consumers and the trade must be made to accept responsibility for managing used products. Different steering instruments must be applied to make sure that those concerned act consistently.

Deposit systems have proved to be the only systems ensuring high percentage returns of used products. Those cases that seem to show a comparable efficiency for voluntary return systems prove, on closer analysis, to be founded on particular circumstances of distribution and tradition.

The deposit can be considered the inducement needed to make the customer take his responsibility for the products after their use. Correspondingly, producers and the trade must be given incentive to accept their responsibility. Direct regulations imposing an obligation on them to take back and to recycle, combined with the possibility of price compensation for the work involved, are a prerequisite for the construction of an efficient recycling operation.

2.6 SWEDISH INITIATIVES

In a Government Bill recently put forward by the Swedish Ministry of Environment and Energy, the aforementioned, integrative approach to waste minimization and pollution prevention is strongly emphasized. Waste is no longer considered as an asset to the community. On the contrary, waste is considered as a burden which has to be reduced by actions taken at the sources of its origin.

Consequently, several administrative, economic and informative steering instruments will be introduced on the market, to force and stimulate producers, importers and consumers to implement waste prevention activities.

The principle of the producers' responsibility to prevent environmental impact, and the Polluter Pays Principle, will be strengthened by the following:

1. the development of economic steering instruments. As an example, environmental surcharges on all mixed wastes (including hazardous), delivered to incinerators and landfills will probably be introduced in the near future;
2. reduction plans for especially environmentally dangerous input materials. These plans will include for example: (*a*) the heavy metals (especially mercury, cadmium, lead, arsenic and their compounds); (*b*) halogenated aliphatic compounds; (*c*) halogenated aromatic compounds; (*d*) chlorinated paraffins; (*e*) brominated aromatic flame-resistant compounds; (*f*) nonylphenoletoxylates; (*g*) phthalates; (*h*) aryl and alkyl-phosphates; (*i*) methylenechloride; (*j*) trichloroethylene;
3. a clarification of the chemical act, that emphasizes the responsibility of producers and importers to substitute dangerous chemical species with less dangerous (the substitution principle);

4. the development of general models for introducing environmental auditing procedures in industry, including the development of material balance-sheets;
5. the introduction of an obligatory environmental product declaration system, and negative environmental labelling systems, emphasizing the environmental hazards of the products.

To reduce the negative impacts on the environment from wastes that will still be produced, the following statements are put forward:

1. source separation methods will be developed and introduced so that incineration and burial of unsorted waste will be terminated by the end of 1993;
2. a deposit system for reusable PET-bottles, with a return rate of minimum 90%, will be introduced during 1991;
3. polyvinyl chloride will be replaced as a packaging material, during the early 1990s.

3

Regulation–legislation

L. Kramer

3.1 INTRODUCTION

3.1.1 The challenge

Discussions on how to legislate on waste have increased in the past decade, despite of the fact that the past 20 years have seen, in all industrialized countries an enormous quantity of waste legislation appear. It seems that not one single country can claim of having resolved its waste problem. Therefore the discussion on how to legislate will continue. Before endeavouring to speculate how legislation on waste could, or should, evolve it might be useful to mention shortly some of the reasons for the present concern.

(a) Changes in the nature of wastes

In contrast to past centuries, which also produced waste, there is one particular 'new' situation which has appeared within the past 200 years, i.e. the great range of new substances produced by the developing chemical industry. Indeed, most of the present waste problems are due to these new chemical substances and preparations derived from them. The European Economic Community (EEC) recently produced EINECS, an inventory of chemical substances which are used within the EEC and found more than 100 000 different substances. The number of different preparations can only be estimated and is thought to be around 500 000.

Many of these substances are bioaccumulative, persistent or toxic or represent a potential threat to the environment. This problem is increased

The Treatment and Handling of Wastes
Edited by A.D. Bradshaw, Sir Richard Southwood and Sir Frederick Warner
Published in 1992 by Chapman & Hall, London, for The Royal Society
UK ISBN 0 412 39390 5, USA ISBN 0 442 31461 2

by the fact that, while some 30 000 substances are thought to be in one way or the other dangerous for man or the environment, the scientific data on these substances are so limited that until now, not more than about 1500 substances have been classified by legislation. Industrial processes have also changed considerably within the past two centuries, leading to new types of products that create new problems for waste elimination, whether this is during production, in the use of the product, or when the production plant is decommissioned. Good examples are nuclear power plants and biotechnological products. In such considerations it must be remembered that all products, at the end of their useful lifetimes, become wastes. As a result new 'wastes' are arising, and the problems of handling of wastes are becoming more complex.

(b) Problems of affluence

These concerns are linked to problems of the affluent society. The number of production units and the number of products have increased dramatically and continue to increase. Urbanization, consumption patterns, transport and other technologies have further contributed to more production, more distribution and more consumption.

This phenomena is worldwide. The numbers of people living in the 'Third World' and the inevitability that their standard of living will improve, suggests that in the third millenium the considerable waste problem will appear in these countries. It is already beginning to show. Despite this, this chapter will concentrate on industrialized, European countries in an attempt to understand the problem and consider developments.

(c) Problems of space and time

While the need for waste elimination is generally recognized, the 'not in my back yard' (NIMBY) attitude has gained importance. NIMBY not only creates particular difficulties for the siting of dangerous waste facilities but also affects the siting of all waste facilities. This widespread attitude contributes to attempts at getting rid of waste in another way, for instance by exporting it to other countries, dumping the waste at sea or by burning it. However, each of these practices has specific disadvantages too.

An illustration of these difficulties is the nuclear waste problem. During the sixties and partly during the seventies, many countries dumped their nuclear waste at sea. This practice was gradually abandoned by the United Kingdom and other countries in 1983. Since then attempts are being made to find sites where nuclear wastes can be stored indefinitely. In most countries, United States, United Kingdom, Switzerland, Germany, Spain, Japan, Italy, the disposal is at present provisional, since it is extremely difficult to obtain acceptance for a site by the local population. The research to find a solution is mainly directed towards finding underground disposal facilities.

There is not only a problem of space but also of time. Organic wastes will normally gradually disintegrate and cease to be a problem. Other wastes, in particular plastics and metals, may last in the environment for hundreds of years. Nuclear waste is considered to have a short- or medium-term lifetime, if radioactivity does not last longer than 300 years. But Plutonium 239, which is also being used, has a lifetime of 24 000 years.

Groundwater that is used on the European continent as a source of drinking water is up to 400 years old. Thus any leaking from a waste disposal site may have long-term effects on the environment. Similar effects can be imagined with soil contamination or sea pollution. Actual deficiences in the safe elimination of waste may thus have long-term effects, which are difficult to assess at present.

(d) Environmental concerns

Environmental policy aims at preserving, protecting and improving the quality of the environment, at contributing towards protecting human health, and at ensuring a prudent and rational utilization of natural resources. In the attempt to achieve these objectives the prevention of harmful substances being emitted into the environment is of paramount importance. In practice this requires the constant, day-to-day attempt to reduce emissions into air, water or soil and to adapt production, distribution and consumption methods to new technologies and techniques. Accidental pollution of the environment is certainly of considerable harm. However, the threat to the environment does not first and foremost come from accidental pollution. It stems from daily, permitted, authorized and accepted contamination of water, air and soil, intensive use of land, economic and recreational activity and so forth.

(e) Waste policy objectives

Against this background, a policy for waste must first of all aim at the prevention of waste generation. It is obvious that the best protection of the environment against waste consists of the development of technologies that prevent existence of waste, even where this leads to some products no longer being produced, e.g. nuclear weapons, or some processes no longer being used, or consumers and users being persuaded to change their lifestyle pattern.

The second aim of a waste policy must be to promote, as far as possible, the reuse and recycling of waste. Indeed, many waste items might be reused, either in the original way, reuse of a bottle, or in a secondary way, such as the extraction of metals from industrial waste or the combustion of waste to produce energy.

Finally, a waste policy must take the necessary steps to ensure that waste is eliminated: tipped, burnt, dumped, without harm to man or the environment, either at present or in the future.

3.2 REGULATORY RESPONSES

(a) Voluntary action is not enough

It would be preferable if these policy aims could be realized by voluntary action from all waste generators. However, this is not realistic, as the sense of responsibility for the protection of fauna and flora, water, air and soil, the landscape, as well as future generations is just underdeveloped in our society. Also, the individual waste generator may not even be aware of the risks which his waste will bear, alone or together with other waste, for the environment. Short-term economic advantage or individual convenience influences attitudes much more than long-term reflections on what might be the best in the general interest of man and the environment.

Legislative action to address the waste problem is therefore necessary and will have to intensify for the foreseeable future. Law making will have to bear in mind the basic considerations that are mentioned above.

(b) The definitions

First of all the subject matter of legislative action has to be considered. Where 'environment' is defined as concerned with living organisms (and man) only, the pollution, which leaking waste disposal sites might cause to underground water, would not even be addressed. A similar conclusion could apply, for instance, to the contamination of the sea by radioactive substances and even the contribution to the destruction of the ozone layer by halogen compounds.

This example of a definition clearly shows the need of extreme precision in legal definitions. Indeed, where a law intends to regulate compensation for environmental damages, a limitation of 'environment' to living organisms is relatively easily conceivable. The problem may be different where pollution prevention or health protection issues are at stake. These seem to require a broader definition of the environment.

The definition of 'waste' is another example. Definitions contain objective, the holder gets rid of it, or subjective elements, the holder wants to get rid of it. However, legal definitions of waste do not take into consideration that derived from the point of view of the environment, it does not matter whether the soil, or the groundwater are polluted from cadmium that comes from waste, from chemical products, from emissions into the air or discharges into the water; any of these 'products' might cause harm. Indeed, it seems that the invention of lawyers to differentiate between emissions, discharges and waste is artificial. It raises from the point of view of environmental protection as many problems as it intends to solve. The need to draft rules that prevent environmental pollution by any form of emission will thus continue to exist.

Similarly the attempts to define 'dangerous waste' and to provide specific rules for its handling, treatment and elimination will need further efforts from the legislator. At present, differentiations between ordinary wastes and dangerous wastes are being made to allow effort to be concentrated on the more specific risks of dangerous wastes. However, the borderline between wastes and hazardous wastes is rather unclear. The origin of the waste might be one criteria, but this source is not always known. The composition of waste is generally not too precisely known either. Finally, the thresholds of when a substance, or a waste, becomes dangerous to the environment are not, and may not be, capable of being fixed with sufficient precision.

As it was stated above, in the area of dangerous substances the EEC list contains more than 100 000 chemical substances which are on the EEC market. Up to now, less than 2000 substances have received a Community wide classification as toxic, corrosive, etc. and appropriate rules for handling, packaging and labelling. One can easily imagine how little is known on substances and preparations considered as 'wastes'.

(c) Action on processes: prevention of waste generation

If we look into the waste production process, we see the situation shown in Fig. 3.1. Legislative intervention will not and does not restrict itself to define the limits of each of the boxes in Fig. 3.1. According to the principles of a waste policy outlined above, it must intervene as regards the different arrows in the figure.

1. The financial incentives for the development and use of cleaner production technologies can play a very important part. Such incentives could also consist of enabling manufacturers to use green labels for their products, provided that these fulfill certain conditions. Other interventions might concern the requirement to use or not to use specific substances. Thus a legal requirement to use recycled glass or paper up to a certain percentage in the production of specific glass or paper products could be imagined. Likewise a requirement to use only biodegradable plastics as packing material for household products could be imagined. On the negative side (prohibition) the use of chlorofluorocarbons (CFCs) is gradually being phased out. The use of asbestos, cadmium and many other substances could gradually be restricted by regulations. Another example could be the replacement of solvent-based paints by water-based paints, promoted through voluntary schemes, financial incentives or prohibitions. Denmark has already prohibited the use of metal cans for liquid beverages. This example might be followed by a ban on aerosols, on non-biodegradable plastics, or on other products.

2. The intervention could also consist of increasing the charge on manufacturers for generating 'non-economic residuals', by taxing the production of

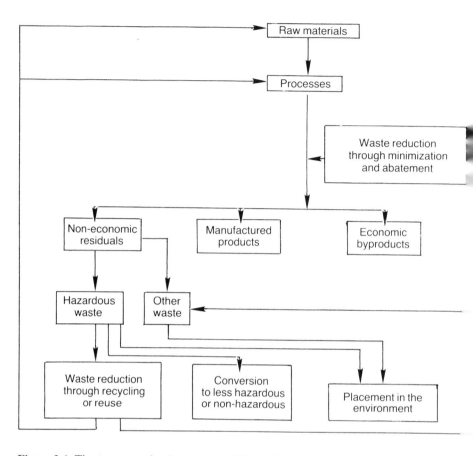

Figure 3.1 The waste production process (NRC, 1985).

waste (and emissions into the environment). This would give a stimulus to the manufacturer to generate less waste. Until now, the general composition of a manufacturer's production is known. Company law and other legal instruments already require, generally, that the turnover and the financial situation of manufacturers be made public, or, where confidentiality problems arise, be at least brought to the knowledge of fiscal administrations. Fiscal inspection takes place to ensure that avoidance of fiscal legislation is made difficult.

No such instruments exist at present by which the environmental outputs of manufacturers are revealed. It is not precisely known what quantity of emissions are sent to the environment in the form of air emissions, discharges into water or into the ground, nor is it known how much waste is generated by a manufacturer. Regulating steps will certainly have to be taken to increase knowledge on these aspects, because only in this way can a better, more preventive environment policy be developed.

Indeed, economic attitudes to waste generation and emissions into the environment are based on the presumption that placements into the environment are free of cost.

'Air and water have traditionally been regarded as "free" goods, but the enormous costs to society of past and present pollution show that they are not free. The environmental costs of economic activity are not encountered until the assimilative capacity of the environment has been exceeded. Beyond that point, they cannot be avoided. They will be paid. The policy question is how and by whom they will be paid, not whether. Basically, there are only two ways. The costs can be "externalized" – that is, transferred to various segments of the community in the form of damage costs to human health, property and ecosystems. Or they can be "internalized" – paid by the enterprise.' (Brundtlandt Commission, 1987, Chapter 8, No. 53)

One of the objectives of environmental regulation must be and will undoubtedly be to ensure that everybody becomes aware that air, water and soil are not free goods.

(d) Re-use and recycling of wastes

Since the generation of waste cannot be altogether prevented, more measures will be necessary to promote the recycling and reuse of waste outputs. Again a huge range of instruments is available to the legislator, from financial incentives to promote new technologies, to the introduction of deposit schemes, fiscal charges for non-recycling, or more directly intervening schemes, such as prohibition to dump at sea, to incinerate or to other forms of placement into the environment. Until now, only part of the potential of reuse and recycling seems to be used in industrial countries, since the placement of waste into the environment is less expensive. This itself is a waste.

The waste strategy paper of the EEC Commission (Commission of the European Communities, 1989) stated that the reuse and recycling of waste should be vigorously promoted, through:

1. research and development on reuse and recycling techniques;
2. optimizing collection and sorting systems, such as selective collection or electro-mechanical sorting;
3. reducing the external costs of reuse and recycling;
4. creating outlets for the products of reuse and recycling.

(e) Disposal of wastes

Legislation will also have to influence the conditions that govern the placement of wastes into the environment. This includes at present, and will continuously do so in future, the physical, chemical, biological or chemical treatment of wastes before land, water or atmosphere are exposed to them.

These legislative measures will have to ensure, among others, that notions such as 'best practicable means' or 'best practicable environmental option' do neither in theory, nor in practice, consider the notion of 'practicable' only or predominantly from the point of view of the waste generator. Indeed, as it was stated before, the use of air, water and land is not free of cost. Societal costs of pollution elimination may be very considerable. The Brundtland report quotes the following estimates for cleaning up waste disposal sites (1986, U.S. dollars) (Brundtland Commission, 1987):

Federal Republic of Germany	U.S. $10 billion
Netherlands	U.S. $1.5 billion
United States	U.S. $20–100 billion
Denmark	U.S. $60 million

The present bargaining between a waste treatment or disposal facility and the administration about what 'practicable' means in concrete terms, all too often lead to neglect of the long-term effects on the environment and the need to take long-term preventive measures. Standardized objective criteria are thus necessary. These might contribute at the same time to assuring the citizen that real efforts are being made to minimize risks. Furthermore, legislative standards help to clarify that the environment cannot be a subject of bargaining between economic agents and administrations, but that society as a whole is concerned with environmental issues. We have not inherited the environment from our ancestors, but borrowed it from our children. If at all, it is thus for the legislator to strike the right balance between the general interest to preserve the environment and the interest of waste generators to place waste in the environment at as little cost as possible.

(f) Accessory measures

Legislation will not only have to address the questions that are directly linked to the generation and elimination of waste. Numerous other regulatory steps will have to be taken and gradually refined to preserve the environment. As regards waste, mention should be made of rules on the transport of waste, in particular the transport to Third World countries. Since waste has little or no economic value the temptation is high to 'lose' it during transport. Furthermore, risks for an accidental pollution of the environment are higher during the transport of waste.

Rules will have to be made on damage caused to man or the environment by waste, be it damage due to an accident or damage caused during an ordinary, authorized handling of the waste.

Accessory rules will have to address the crucial problem of access to information where local authorities do not even know what treatment and elimination standards are being applied in waste treatment plants or waste facilities, which operate on their ground. There can be no doubt that emotions are high at the local level and that NIMBY arguments are frequent. An open society, which the society of the next millenium hopefully will be, can simply not afford to keep data on standards applied and emissions occurring secret.

Similar considerations will have to apply to general questions of monitoring and enforcing existing rules. Wastes have a long-term risk potential for man and the environment. Waste disposal sites and other waste treatment facilities therefore need a permanent surveillance system, which is capable of discovering leakages to the environment, provide and enforce repair work, and provide, where necessary, for the adaptation of risk prevention technologies to new knowledge, new methods and new techniques. Thus there is a permanent task for waste monitoring, inspection and surveillance of waste facilities, which requires permanent adaptation of legislative standards to scientific and technical progress.

3.3 CONCLUSIONS

'Pollution is a form of waste and a symptom of inefficiency in industrial production' (Brundtland Commission, 1987, chapter 8, No. 51). Waste is a form of pollution and a sympton of inefficiency in industrial production. The less waste that is generated, the more efficient an economic system is. The prevention of waste generation, the promotion of reuse and recycling of waste and the safe elimination of waste, as well as the need to minimize emissions into the environment are permanent tasks. They will increase since economic production will increase and more and more countries accede to the benefits of affluent societies. The legislator of the Third Millenium will therefore have to set up rules that are permanently adapted, monitored and enforced to ensure that the society is not destroying its own environment.

REFERENCES

Brundtland Commission (1987) *Our common future*. World Commission on Environment and Development, Stockholm.
Commission of the European Communities (1989) A Community Strategy for Waste Management, SEC (89) 934. Commission of the European Communities, Brussels.
Natural Research Council (1985) *Reducing hazardous waste generation*. National Research Council, Washington.

Part Two

Transformation and Re-use

Chairman's introduction

C.W. Suckling

The subject of this session is so wide that it is impossible to overview the many underlying problems and opportunities in a short space. What needs to be said is that transformation and reuse, by which wastes are no longer wastes, should be the goal of all industrial processes. As will be seen from the papers of later sessions, all the alternative approaches have problems, because it is likely, no matter what technology, there will always be limitations to what can be achieved. So the design of industrial processes to reuse wastes or, if possible, to eliminate waste production in the first place, should be the universal aim. It is appropriate that this is the approach advocated in the final contribution to this meeting. The economics of reuse have to be understood. It is easy to take a narrow, least possible cost, approach to waste disposal. But when wider, social cost-benefit analysis is undertaken, the balance of what is worthwhile changes, and reuse, so easily dismissed in a short-sighted approach, becomes economically justifiable. The analysis by Professor Turner is an important contribution to this meeting.

But all this will depend on the practical realities of the nature of the waste. Never is this more clear than in the example of sewage sludge. The potential benefits of sewage sludge are widely appreciated, but reuse is severely hampered because of the content of pathogens and, more particularly, toxic metals. The former can be dealt with, and new technology will help, but the latter cannot. There is only one answer, to change our sewage systems to segregate toxic metals at source, and, where this is not possible, to restrict their use. This is a radical and difficult but not impossible approach. In the end the best solution to our waste problems will be to design products and processes backwards, beginning with their final impact on the environment.

This theme was very much in the mind of the Royal Commission on Environmental Pollution when we wrote our 11th report in 1985 on *Managing waste – the duty of care*. I refer to the chapter entitled Professionalism, design, standards and training – without which the potential benefits of advanced technology will not be realised.

We defined professionalism as the skilled and conscientious performance of a task, a definition which has no elitist limitation.

As this conference shows, the level of professionalism in many parts of the waste management industry is high. But we would, I suspect, agree that, at least in respect of skills, this is not always the case. Moreover, one's professional skills need continual reinforcement, continual topping up if they are not to become outdated.

The Royal Commission coupled professionalism with design. Designers in any field need to be aware not only of the technical and economic demands at all stages of the life of their process or product, for as long as it might present a risk to human health or to the environment, but also of the changes that manufacturers, operators, users and the public at large will need to make to use and to benefit from the new design.

This is clearly important in the waste treatment and waste disposal industry.

There is widespread evidence on the continent as well as in the U.K. that people are willing to pay to protect the environment, but are extremely loath to have any manifestation of the need in their own locality, the 'not in my back yard' (NIMBY) syndrome is universal; none of us wants an incinerator or a waste collection unit near where we live, let alone in our own back yard.

As an essential element in process design, there should be an intensive and imaginative review of the plant as it will operate in the intended location, by a technique such as hazard and operability study (HAZOP), which systematically investigates the hazards that might arise if a process operates *outside* the design intention, and then considers whether the resultant risks demand further precautionary measures.

It is sad that we so often rely on the benefit of hindsight, as expressed, for example, by a judicial enquiry after a disaster, when most, if not all, the hazards could have been foreseen by the application of a technique such as HAZOP. Of course, the judicial enquiry does serve the important purpose of putting considerable weight behind the implementation of its recommendations and giving wide publicity to them. But hopefully, with better management commitment and a professional approach to design we can operate more often with the benefit of foresight.

It is sometimes said that procedures such as HAZOP are too expensive and too time consuming, and that in any case an experienced designer knows already what the hazards are and how to deal with them. But no plant is an exact copy of an existing one, and it will often be operated in a new location. HAZOP studies and the like do demand a lot of effort, but the payback in terms of increased safety and reduced environmental impact not to mention better start-up and smoother operation can be enormous.

The definition of a design competence offered by the designer Norman

Potter is a useful checklist of the elements that a professional in any field should bring to bear on his tasks:

'A design capability proceeds from a fusion of skills, knowledge, understanding and imagination; consolidated by experience.' or substituting 'professional' for 'design':

'A professional capability proceeds from a fusion of skills, knowledge, understanding and imagination; consolidated by experience.'

Try it out, substituting your own discipline for design.

Consolidated by experience; experience can also be a motivator when our eyes are opened by seeing what has been achieved elsewhere, extending our perception of what is possible. Experience of advanced science and technology in the relevant fields and of the best management practices is not always available to those in the waste management industry, or indeed in those industries that create waste. It is, however, an important element in ongoing education, in the training that was referred to by the Royal Commission.

And as for training, is it sensible that anyone, without appropriate training and qualifications can legally manage or operate plants in the chemical and waste management industries?

Even the best designed plant needs conscientious managers and operators. A critical element in conscientiousness in the manager is to know what is actually happening within his or her sphere of responsibility. For example, from time-to-time to walk the plants, not to snoop, but to support while noting such things as whether prescribed procedures for maintenance are being followed, whether safety equipment is in place and alarms and other protective devices are in fact in operation.

Coming from the chemical industry as I do, I am very much aware of the need to design all processes to avoid waste as far as is possible and I am very glad to see that the first paper in this session will be a contribution by 3M on waste management by prevention. This necessity also was a principal theme of the Royal Commission's Report on Waste, to which 3M made a valuable contribution. It was refinforced in the 12th Report of the Royal Commission on Best Practicable Environmental Option.

In conclusion, I would like to mention just one area of design which offers great promise for the avoidance of pollution of all kinds, process intensification: by which I mean finding ways in which one can achieve a given transformation with far less equipment, increasing the space time yield of as many units as possible.

Research in chemical companies such as ICI has shown that it is possible with some processes, once the limiting factors are recognized through elucidation of the kinetics, to bring about transformations in a much simpler way and in much smaller units than had previously been thought necessary.

The raw material and energy needed to manufacture plant items is reduced. There is less material in process at one time. Processes can be made more selective. Smaller plant will create less visual intrusion in the environment. Moreover it is possible in some instances, as for example in electrolytic cells, to design a small compact unit which can be economically operated at the sites in which the chemicals are needed.

Work such as that of Professor Bodo Linnhof on maximizing the use of driving forces, such as temperature or concentration differences, and successively eliminating pinch points also make an important contribution, both to energy efficiency and to the reduction in size and complexity of the plant required. All this results in reduced actual and potential pollution and increased safety, as well as in reduction in fixed and working capital.

These possibilities are enlarged by the vastly improved techniques for studying the mechanisms of chemical processes, which should, for example, lead to the design of catalysts with better selectivity, greater efficiency and longer life. For other processes a study of kinetics has led to the recognition that much improved throughput and purity could be achieved by more effective mixing, for example in a venturi. Other studies have led to the elimination of process stages, for example by ensuring that a product is precipitated in usable form, not requiring subsequent comminution. And so on.

These initiatives together with the epoch making opportunities afforded by the use of micro-processors in sensors, transducers and in control have opened up a new field for environmental friendly plants that is only beginning to be explored.

Given an understanding of the underlying science, a professional commitment to design, to construction and to operation of our plants we could, with the benefit of foresight, achieve in this millennium important advances in reducing the production of waste and in managing what remains unavoidable, as well as providing a springboard for the revolutionary progress in the early decades of the next millennium that will undoubtedly be needed to secure an environment that is fit for all to live in.

4

Waste management by prevention

R.P. Bringer

4.1 INTRODUCTION

In every instance today, where an industrial corporation creates a waste, it is increasing its level of corporate risk and decreasing its ability to compete. In the future, that cause and effect relationship will only get stronger.

At 3M, we believe one of the keys to competing in the 1990s and beyond is a constant programme of waste reduction through preventive measures. Non-polluting products and processes will be the hallmarks of leading corporations.

4.2 WASTE MANAGEMENT AS A COMPETITIVENESS ISSUE

Historically, waste has always been part of our industrial processes, but mostly viewed as an undesirable by-product. Because of that view, waste management in the past consisted of getting rid of waste in the quickest and cheapest way possible. This often meant taking advantage of 'free' disposal into the air, water and land. In essence, waste management was a non-issue.

Over the past 15–20 years, waste management has become more of a regulatory compliance issue as a substantial number of laws have been passed to protect the environment.

Corporations now accept this as a cost of doing business and have organized to deal with environmental regulations. In more recent years, continued regulatory pressure has combined with public opinion to place additional constraints on how corporations conduct their businesses.

Not only is our manufacturing flexibility being limited today, but our very

The Treatment and Handling of Wastes
Edited by A.D. Bradshaw, Sir Richard Southwood and Sir Frederick Warner
Published in 1992 by Chapman & Hall, London, for The Royal Society
UK ISBN 0 412 39390 5, USA ISBN 0 442 31461 2

ability to manufacture in certain locations, to use certain materials and to sell certain products. Not all of these restrictions are forced. Some are being placed on corporations by themselves.

The number and restrictiveness of regulations continues to grow at the national, local and now even the international level. In the United States federal regulations are estimated to have grown 25% per year over the past 10–15 years. The direct costs to industry of coping with those regulations has also continued to grow at much the same rate. Those costs have now reached significant levels in the corporate economy. This has meant a siphoning of resources from certain discretionary activities, such as research and development or productivity improvement.

The future outlook is for even more regulation. Environmental groups, who normally favour increased regulation, have been growing in numbers, resources and political influence. The media are influential in their instantaneous and graphic depiction of frequently occurring industrial mishaps. And people, in general, are giving more importance to personal health effects. They are also beginning to see and hear about environmental problems that are affecting all individuals, such as municipal waste, ozone depletion and global warming.

In addition to the regulatory outlook, there are new market forces at play today, which are causing corporations to rethink the waste management issue. Forward-looking corporations are recognizing that waste management is going to be a competitiveness issue in the 1990s and that waste reduction will be the key to success. Waste reduction in both our processes and our products offers the only hope of coping with the regulatory, litigation and public opinion whirlpool which is steadily increasing in intensity. In addition, there are cost, quality and marketing benefits to be had.

In today's global competition, corporations are finding it necessary to compete broadly in their markets. That means they must not only sell the high performance, high profit products, but also the lower performance, lower profits products. To be able to do the latter well, it is important to be a low-cost producer. High waste and low cost are usually incompatible.

Another important aspect of today's global competition is quality, both in our products and in our processes. Again, high waste and high quality are usually incompatible.

Products that do not create environmental problems during manufacture or use are becoming more popular with consumers. Germany and Canada already have programmes to identify products that are environmentally benign. Other countries are looking at similar programmes. And it is not just the public consumer who is showing this interest in low or non-polluting products. 3M, whose sales are only 10% direct to consumers and 90% to other companies, has been getting many strong suggestions from our corporate consumers that they would also prefer low or non-polluting industrial products.

The message on being competitive in the 1990s and beyond is clear. One must be a high quality, low-cost producer of low waste products using low waste processes. The encouraging part is that these goals are all compatible.

4.3 WASTE MANAGEMENT AS A RISK ISSUE

Environmental risk management in an industrial corporation is a fairly recent phenomenon, and not a very well understood one. Risk managers in industry have been associated in the past with insurance departments and the recovery of financial losses in cases of unforeseen disasters. The fact is that corporations are built and managed broadly on the whole concept of risk. Each business decision, no matter how large or small, is based on the degree of risk of success (or failure) and the amount of investment.

Generally, the higher the investment involved, the lower risk of failure is tolerated. And so we get to the general definition of risk:

the *probability* of some event happening × the *seriousness* of the consequences if that event does happen.

Environmental risk is limited to situations in which an event can have environmental degradation and/or human health effects. Environmental risk management then consists of three basic elements:

1. risk assessment, which is the determination of the probability of any given environmental event happening and the seriousness of that event;
2. a risk management decision which either classifies the risk as acceptable or in need of reduction;
3. risk reduction, which can be carried out on either the probability aspect, the seriousness aspect, or both.

Corporate environmental risks can be associated with either facilities or with products. Risks can also be associated with unexpected events, like industrial accidents, or with routine events, such as permitted air and water emissions.

The number of environmentally related events that could happen accidentally or without foreknowledge, or that do happen routinely, can be quite large for a multinational industrial corporation such as 3M. Put together 150 manufacturing plants and 50 000 products with 85 000 people making decisions effecting those plants and products, and the resulting number of potential events makes it virtually impossible to carry out a risk assessment for each. As a result, risk assessments are carried out mainly in situations where the volume and toxicity of the materials involved merit the effort and cost for what is often an imperfect exercise.

So how does a corporation manage its environmental risks with incomplete assessment of all its risk situations and imperfect knowledge in those cases where it has assessed risk? At 3M, we have taken the position that

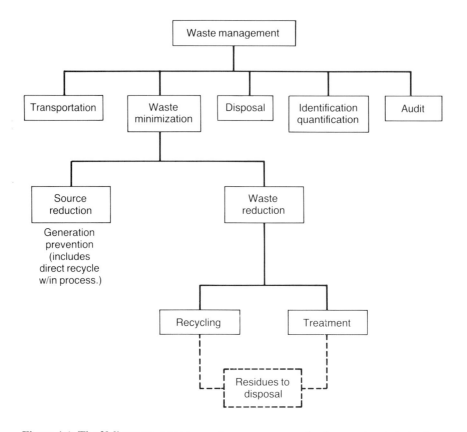

Figure 4.1 The 3M's waste management programme emphasizes waste minimization.

environmental risk management really equates to environmental risk reduction which, in turn, equates to reduction of waste generated and emissions to all media. We have taken this position because the risk assessment process is imperfect and too burdensome to complete. In addition, there is the public's perception that risks need to be reduced in any event. There are also the more generalized risks we will encounter if we do not follow a policy of waste or emission reduction in both our plants and our products. Some of those more generalized risks are:

1. non-compliance, fines and penalties can be substantial;
2. generation of future liabilities, any emission, legal or not, can be a cause of future litigation;
3. loss of community support, availability of emissions data to the public and the perception that volume equates to health risk;
4. loss of reputation, could be accompanied by a loss of business;

5. loss of employee support; if they perceive you are not interested in their safety and health;
6. direct loss of business; if competitors have products with lower environmental risks attached;
7. becoming non-competitive; waste not only affects the cost of production but also the quality of our products.

4.4 3M's WASTE MANAGEMENT PROGRAMME: PAST AND FUTURE

3M's Waste Management Programme emphasizes waste minimization today (Fig. 4.1) as it has since 1975. In fact, Waste Minimization is the cornerstone of our Waste Management Programme and events since 1975 have only served to reinforce our conviction that that is where it belongs. In 1975, we developed an Environmental Policy (Fig. 4.2), which still stands today. It emphasized waste reduction at the source and led to the formation of our

Solve own environmental problems

Prevent pollution at the source

Conserve natural resources

Develop environmentally sound products

Comply with all regulations

Assist official environmental organizations

Figure 4.2 3M environmental policy.

corporate waste reduction programme called Pollution Prevention Pays (3P). The 3P Programme is a voluntary global programme which encourages our employees to carry out projects that not only have an investment return in line with company expectations, but also have a pollution prevention or environmental return. Acceptable projects are catalogued and suitably recognized by corporate management. Over the past 14 years, the 3P Programme has accepted 2444 projects which have generated U.S.$482 million in cumulative first-year only savings (Fig. 4.3) and prevented approximately 500000 tons of pollution annually (Fig. 4.4).

The 3P Programme was a model for today's widely accepted Waste

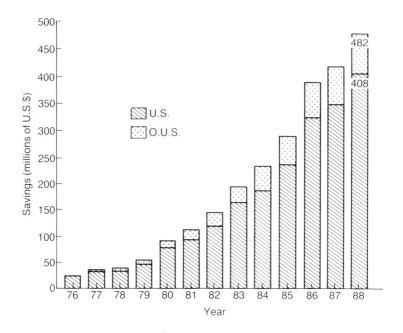

Figure 4.3 3P Programme global savings for 1976–1988 (cumulative).

	United States	International
Air pollutants	111000 Tons	111000 Tons
Water pollutants	15 Tons	1100 Tons
Waste water	1 Billion Gallons	600 Million Gallons
Sludge/solid waste	388000 Tons	12000 Tons

Figure 4.4 3P global results for 1975–1988, pollution prevented annually.

Management Hierarchy (Figure 4.5). It says that, from an environmental standpoint, source prevention and recycling are preferred over treatment and disposal. What the 3P Programme has evidenced is that, in many cases, the Waste Management Hierarchy not only works environmentally, but economically as well.

Although the 3P Programme has economically reduced our waste generation to about 50% from what it might have been, a great deal of waste is still

Waste Management Hierarchy

Source reduction

Recycle/reuse

Treatment

Disposal

Figure 4.5 The Waste Management Hierarchy, where source prevention and recycling are preferred over treatment and disposal.

generated. Much of this is in the form of air emissions of hydrocarbon solvents from our many coating processes. In 1987, corporate management decided, for a number of reasons, that the 3P approach was not reducing our air emissions fast enough. We instituted a five-year global plan to put controls on all our major air emission sources at a total cost of U.S.$150 million. This plan is on schedule. It will be completed in the U.S. by mid-1992 and globally by mid-1993. It will reduce our overall air emissions by about 70%. This plan was initially conceived as using pollution control equipment to reduce air emissions, and thus not utilize the 3P approach. However, it has had two effects which encourage the 3P or prevention approach. First, some operations which would be forced to add expensive controls found a way to accelerate research to eliminate solvent use. Second, those operations which have to install controls recognize the costs of operation, maintenance and replacement and will work hard to eliminate their need in the future.

This year, 3M is embarking on a new waste reduction programme which we are calling 3P Plus. It will continue the principles of the 3P Programme, but add corporate goals and more structure to its implementation.

The goals of the 3P Plus Programme are shown in Figure 4.6. They include quantitative goals for multi-media emissions reduction and waste generation. There are also non-quantitative goals encouraging the development of products with low waste potential and encouraging our suppliers to do the same. These goals will be translated into individual operating division and plant goals.

On the implementation side, more assistance will be given to the operating units. Waste streams have already been identified. We will try to relate them to specific products and to assess full (real and intangible) costs for

3P Plus Goals
_____

Reduce emissions 90% by year 2000
long term goal:
    emission-free operations

Reduce generation of waste 50% by year 2000

Build waste prevention into products wherever feasible

Demand waste prevention from suppliers

Figure 4.6 Goals of the 3P Plus Programme.

ultimate waste handling and disposal. This should generate economic incentives and allow priority setting. We will identify opportunities for waste reduction. Corporate staff teams will be available to assist in implementing the actual waste reduction projects. Progress against goals will be tracked and used as a recognition tool.

In 1991, the 3P Plus Programme will be phased into the corporate Manufacturing Competitiveness (MC) Programme. The present five-year MC Programme which runs through 1990, has reduction goals for direct labour cost, cost of quality, and product cycle time. The 1991 MC Programme will also last five years and have three goals, one of which will be the 3P Plus goal of waste generation reduction. This will be corporate management's way of saying 'Waste reduction/management is now a competitiveness issue.'

4.5 THE NEED FOR A LONG-RANGE PHILOSOPHY

Industry needs to realize that in the long run, non-polluting processes will be the most economic, allow the greatest operating flexibility, and produce the highest quality products. In the long run, non-polluting products will be the most profitable and most marketable. In other words, such products will be most competitive. This is important in today's global competition, essential in tomorrow's.

Industry needs to recognize the long-term aspect of a move toward non-polluting products and processes and the need for research and development expenditure. Some significant share of today's spending must go toward tomorrow's non-pollution goal. And that goal has to be built in to all forms of research and development, long range or short.

And industry needs to encourage a pollution prevention philosophy and link it to economic goals. It's the only way we can ultimately meet the competition, reduce our environmental risks to minimal levels, and obviate the need for ever-increasing regulation.

5

Treatment and use of sewage sludge

J.E. Hall

5.1 INTRODUCTION

5.1.1 Historical perspective

The building of sewerage systems in the developing cities of the eighteenth and nineteenth centuries resulted in the greatest single improvement in public health. Cholera, typhoid and other enteric diseases were common among urban populations but are now absent from developed countries. However, the water carriage of human wastes produces large volumes of sewage, which require purification. Coastal towns have commonly discharged sewage directly into the sea, while inland sewage was at one time applied to sewage farms where the soil served to filter and transform the polluting load. Land treatment was replaced by methods that concentrated the polluting load into a small volume, which could be treated further before final disposal, whilst the bulk of the water after purification could be returned safely to watercourses. As the need developed for the reuse of water from rivers, higher standards of effluent treatment were required and consequently more sewage sludge was produced.

Sewage sludge requires safe and economical disposal, and recycling to farmland has long been the natural and most important method. Its fertilizer value was quantified as early as 1859 by Mechi: '200 tons of London sewage is equivalent to 3½ cwt of guano'. Hibberd (1863) also rated its value higher than any other manure. Sludge contains similar quantities of nitrogen, phosphorus and organic matter as farmyard manure or slurry, making sludge an attractive supplement to other fertilizers, particularly in predominantly

The Treatment and Handling of Wastes
Edited by A.D. Bradshaw, Sir Richard Southwood and Sir Frederick Warner
Published in 1992 by Chapman & Hall, London, for The Royal Society
UK ISBN 0 412 39390 5, USA ISBN 0 442 31461 2

arable areas. Sludge may also contain lime and other essential trace elements but little potassium. The potential savings in fertilizer provided by sludge spread on farmland each year are in excess of £15 million. However, there are also constraints on the recycling of sludge because of heavy metals and pathogens that may be present, the occurrence of which reflect the nature of the catchment of the sewage treatment works (i.e. industrial, domestic or mixed). Furthermore, there are the potential problems of odour and bulk (sludge may contain >95% water). While there have been considerable advances in technology in recent years, which have reduced environmental impact and costs, sludge quality remains one of the principle constraints on the recycling of sludge.

Although agriculture remains the most important outlet for sludge, other options have been developed; some of these, such as landfill and the sea are merely disposal options, whereas others such as land reclamation and forestry are beneficially affected by sludge addition. The choice of disposal outlet has in recent years been governed by the concept of 'best practicable environmental option' and is a balance between sludge quality, treatment and disposal costs, security of outlet and environmental impact.

5.1.2 Future prospects

With our present knowledge, there seems little prospect of developing a cost-effective sewage treatment process which does not transfer a significant proportion of the pollutant load into a concentrated wet solids sidestream requiring off-site disposal. For the foreseeable future, therefore, sewage works will continue to function as 'sludge factories' with unceasing and unstoppable output. Sewage sludge will also remain a product the quality of which is not strictly controllable, which may have no secure long-term outlet and which usually entails processing, transport and disposal costs of up to half the total cost of operating the sewage works. Sewage sludge is thus often regarded as the major problem of water pollution control technology.

The Member States of the European Community (EC) currently operate over 30000 sewage treatment works yielding a total of some 5.5 million tonnes of dry solids (tds) per year (Bowden, 1987). With further major water pollution control schemes in prospect, EC sludge production is expected to rise by at least one third over the next 10 years. In the U.K. where over 85% of the population is already served by sewage works, sludge production is just over 1 million tds per year and will continue to rise gradually. U.K. sludge treatment and disposal costs are now of the order of £250 million per year.

The disposal of sewage sludge always requires very positive and careful management, but the ease, or difficulty, with which disposal is actually achieved, and the associated costs will obviously depend very much on circumstances. Both local and national geographical, agronomic, economic and 'political' factors will have some influence. The general trend in recent years in most developed countries has been for the disposal of sewage sludge to become more, rather than less difficult. The problem applies particularly to the more densely populated areas; for example, in the Netherlands where the agricultural outlet for sludge is becoming increasingly restricted and is competing with increasing quantities of animal manures (de Bekker & van den Berg, 1987). In Switzerland, agricultural use of sludge has decreased markedly during the last few years because of stricter requirements. Land-fill represents another option but restrictions on this outlet are increasing (Lichensteiger *et al.*, 1987). Sea disposal is not an option for most countries except the U.K. which accounts for nearly 30% of its sludge production but this practice will be phased out by December 1998. Incineration is seen by many as the only feasible option in the face of such increasing difficulties but public opposition may make this difficult to implement.

All EC states are now subject to a new Community Directive designed to protect the agricultural environment from adverse effects of sludge utilization on land (Council of the European Communities, 1986). This imposes further restrictions on sludge disposal though the general effect will be to promote better control of sludge quality and use to help ensure that the agricultural outlet is both safe and beneficial. There are a number of other EC Directives, still at the proposal stage, on waste water treatment and nitrate emissions which will tend to increase the amount of sludge for disposal, increase treatment costs and restrict sludge disposal options.

In recent years, sewage sludge has become an international topic with numerous conferences and, in the case of the EC, interstate coordinated research and scientific committees focusing on various common problems (Commission of the European Communities, 1983). This activity reflects the growing realization that while world sludge production is on a relentless growth curve, environmental quality requirements for sludge are becoming increasingly stringent, disposal outlets are decreasing and yet economic pressures still require low-cost solutions to sludge disposal problems.

5.2 SLUDGE PROCESSING

5.2.1 Sludge quantity and quality

There is now fairly good agreement on rates of production of different types of sludge so predictions can be reasonably accurate for process design purposes (Poulanne, 1984). By using conventional sewage treatment technology, it is difficult to achieve any substantial reduction in production rates. However, the search for a method to significantly reduce production will continue. The wet oxidation of sludge solids is one possibility and has been examined in a number of studies (Fisher, 1971). This process represents an almost 'sludgeless' method of sewage treatment but practical considerations and costs rule it out, except in very special cirucmstances. In the U.K., a process known as 'LOSLUJ' is under investigation (Hoyland & Ronald, 1984) which produces solids at a rate about 40% lower than for conventional treatment and reduces operating costs by approximately 20%. However, the prospects of employing more than a few of these plants in the next few years are limited, so they will have a minor effect on national sludge production.

The quality of raw sludge is only partly under the control of sewage works management. By using trade waste control, the concentrations of potentially toxic components can be reduced but not eliminated, and great improvements in this direction have been seen in the U.K. and other countries in recent years. Domestic sludges represent the least contaminated type, but some heavy metals, detergents and other organic residues are always present. Overall, only about 20% of the heavy metal contamination of sewage sludge is derived from industrial inputs. Table 1 gives the ranges of concentrations of the significant potentially toxic elements which occur in sludge.

Economic considerations rule out the use of special processes at sewage works to remove chemical contaminants from sludge. Although the extraction of heavy metals is technically feasible, the contaminants would still require disposal somewhere.

The pathogen content of raw sludge is determined basically by the health of the local contributing population and cannot be controlled initially. However, techniques of disinfection are now available and a great deal more is known about the effectiveness of conventional processes in destroying pathogens.

A further potential problem with regard to the quality of sludge is its content of rag and other litter material. This can present a problem when sludge is spread on land or in the marine environment. In recent years, advances have been made in the fine screening of sewage and successful methods of sludge screening are now also available. The practice of fine

Table 5.1 Concentrations of potentially toxic elements in sludge (mg kg^{-1})

PTE	Range	Average	Domestic
Cd	2–1500	20	5
Cu	200–8000	650	380
Ni	20–5000	100	30
Zn	600–20000	1500	515
Pb	50–3600	400	120
Hg	0.2–18	5	1.5
Cr	40–14000	400	50
Mo	1–40	6	4
As	3–30	20	3
Se	1–10	3	2
B	15–1000	50	50
F	60–40000	250	200

screening is a growing one and its beneficial effects are likely to become
more widely appreciated when seen in the context of improved ease of
treatment and disposal (Holladay, 1986).

5.2.2 Sludge treatment in relation to disposal

It can be argued that selection of the best disposal route for sludge from a
particular works should start by identifying the most secure and environmentally acceptable final destination for the sludge and this, in turn, will
dictate the type of treatment required before disposal. In theory, at least,
the sludge processing requirements might even determine the type of sewage
treatment process employed. For example, a high-rate activated sludge
process is not ideal where dewatering to a high-solids cake is required. This
reverse sequence of selection procedures rarely occurs in practice and,
indeed, sludge disposal has often been done on an ad hoc basis with each
work's management determining a local sludge disposal solution, taking into
account such factors as historical investment, sludge quality and disposal
outlets. However, in recent years in the U.K., sludge disposal strategies for
much larger areas (e.g. river basins) have been evolved.

The current options potentially available for sludge disposal in the U.K.
are shown in Fig. 5.1. These are:

1. agriculture (including horticulture);
2. other on-land outlets;
3. the marine environment;
4. specialized outlets such as building materials, feedstuffs, fuels, etc.

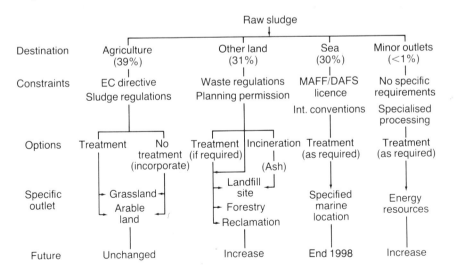

Figure 5.1 Options and constraints for sludge disposal.

It can be seen that specific outlets may be reached only by complying with certain constraints and these are likely to change in the future as environmental and legislative constraints increase.

The most basic end product, raw sludge (thickened or unthickened), can still represent the most economic type for disposal provided a suitable outlet is available. Raw sludge is becoming less and less acceptable environmentally, but even under the EC Directive it can still be used in agriculture provided it is immediately injected or ploughed into the soil. Raw sludge may also go to sea (until 1998) but there is a need to ensure that it does not cause visual pollution from grease and floating solids.

For most situations, however, sludge needs first to be treated. In the EC Directive, 'treated sludge' is defined as 'sludge which has undergone biological, chemical or heat treatment, long-term storage or any other appropriate process so as significantly to reduce its fermentability and the health hazards resulting from its use'. This is clearly a rather loose definition of treatment, but the general intention is clear, that is to process sludge so as to reduce its nuisance value from odour and to reduce significantly the number of pathogens present. Individual EC states provide closer definitions of the treatment actually required within their borders in order to meet the requirements of the Directive. In the U.K., a Code of Practice indicates a range of treatment processes which will be deemed to satisfy the definition of 'treated' as given in the Directive (DoE, 1989).

The requirements for treatment before landfilling or land reclamation will depend on the particular site and the site owner's requirements. The general trend for landfill is to require sludge cake which is physically stable. This may require dewatering to a solids content of well above 30%, so appropriate dewatering systems have to be used.

For marine disposal, the conditions of the government licence will specify maximum quantities and maximum amounts of specific pollutants that may be disposed of at sea, but there may be no definite requirement for treatment. However, it is often in the best interests of the producer to, for example, thicken the sludge before disposal to reduce volume and to stabilize, by say digestion, to minimize the quantity of sludge solids for disposal and to control local odour nuisance.

The range of unit operations and processes which may be used to modify the quantity and character of raw sludge before disposal are summarized in Fig. 5.2. Essentially, there are six which can be employed, either alone or in combination.

1. screening to remove gross solids;
2. thickening to reduce liquid volume;
3. positive disinfection to destroy pathogens;
4. stabilization to improve odour and reduce pathogens;
5. dewatering to form a solid cake;

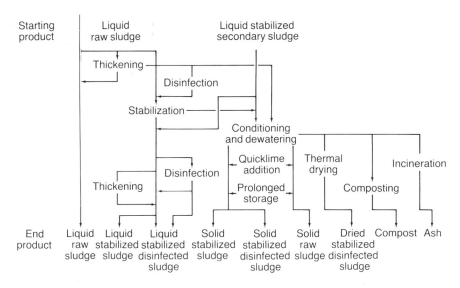

Figure 5.2 Flow chart of sludge processing options for production of suitable end-products for utilization or disposal.

6. thermal destruction of organics (incineration) to minimize the mass of solids for disposal.

Specialized treatment/conversion processes such as pyrolysis, protein extraction, etc., are not included as they are used only in rare circumstances, and not at all in the U.K. as yet.

5.3 SLUDGE RECYCLING/DISPOSAL OPTIONS AND CONSTRAINTS

5.3.1 Agriculture

The spreading of sludge on agricultural land recycles plant nutrients, principally nitrogen and phosphorus, and organic matter. The fertilizer values of the different types of sludge that are offered to farmers are now well recognized (WRC, 1985) and are summarized in Table 5.2. Application rates are limited to supply up to crop requirements for nutrients and are normally in the range 50–100 t ha^{-1}.

Techniques of sludge application have improved considerably in recent years to ensure that correct and even application is achieved, odour and visual nuisance is avoided and minimal damage is done to the farmer's land (Hall 1988, 1989; WRC 1989). By and large, sludge is supplied, and often spread, free of charge, but nevertheless the sludge disposal authorities rely very heavily on the goodwill of farmers for continuity of the outlet. The use

Table 5.2 Typical sludge analysis (WRC, 1985)

Sludge type	Dry solids Content (%)	Nitrogen Total (%ds)	Availa- bility (%)[a]	Phosphate Total (%ds)	Avail- bility (%)[a]	Nitrogen form
Liquid undigested	5	3.5	35	3	50	largely organic
Liquid digested	4	5	60 (60+40)[b] (100+15)[b]	4	50	largely ammoniacal
Undigested cake	25	3	20	2.5	50	organic
Digested cake	25	3	15	3.5	50	organic

[a] Availability in the first year.
[b] Ammonia + organic nitrogen.

of sludge in agriculture is well regulated so as to avoid potential problems from potentially toxic elements (PTEs) and pathogens. The Sludge Regulations (Statutory Instruments 1989) is the implementation of the EC Directive (CEC, 1986) but there is also a Code of Practice (DoE, 1989) which includes additional non-statutory control measures; this replaces the earlier DoE guidelines (DoE/NWC, 1981) but the overall objectives remain the same in ensuring that:

1. there is no conflict with good agricultural practice;
2. the long-term viability of agricultural activities is maintained;
3. public nuisance and water pollution are avoided;
4. human, animal or plant health is not put at risk.

(a) Grassland

This is the main form of agriculture to the north and west of the U.K., in Ireland and in upland regions of Europe. For sludge utilization, grassland has the advantage that it is likely to be accessible throughout the year.

It has the disadvantage that there is a risk of disease transmission to grazing animals because of the possible presence in the sludge of pathogens such as *Salmonella* or beef tapeworm eggs (*Taenia saginata*). If sludge is spread on the surface of grassland, it is necessary first to treat it by a process that significantly reduces numbers of pathogens so as to avoid disease transmission (DoE, 1989).

The EC Directive does not permit the application of untreated sludge to grassland unless injection is used to bury the sludge beneath the soil surface. Where soil conditions and operator skill result in successful injection, there is the dual advantage of avoiding both pathogen risk and odour nuisance. This can otherwise be a major source of complaint from the public in built up areas. Grassland is best suited to receive dressings of liquid-digested sludge because digestion gives adequate control of pathogens and the sludge contains sufficient plant-available ammoniacal nitrogen to sustain rapid growth of grass. Dewatered sludge, unless applied thinly and evenly, can leave persistent lumps of sludge on the surface of the grass which may cause sward damage or adversely affect grazing animals if ingested.

As with other land utilization outlets, sludge applied to grassland is subject to soil limits (Table 5.3), and because grazing animals inadvertently ingest soil, there are special limits on lead and fluorids concentrations in sludge of 1200 and 1000 mg kg^{-1} ds respectively to protect animal health.

Surface applications of sludge to grassland lead to accumulation of PTEs in the surface layer of the soil profile (Davis *et al.*, 1988), and thus higher PTE soil limits are permitted provided that the depth of sampling is to 7.5 cm compared with the statutory sampling depth of 25 cm. Permanent pasture remains unploughed mainly for reasons of soil conditions, topography and climatic factors which may also make such land unsuitable for sludge spreading. Permanent grassland soils also tend to be acidic and this requires working to lower PTE limits because the availability of certain PTEs increases as pH falls (Table 5.3). No sludge can be applied to soil where the pH is less than 5.0.

Table 5.3 Maximum permissible concentrations of PTEs in soil (mg kg^{-1}) and maximum permissible average annual rate of PTE addition over 10-year period (kg ha^{-1}).

	Soil limits (mg/kg)				
PTE	*pH 5.0<5.5*	*pH 5.5<6.0*	*pH 6.0–7.0*	*pH >7.0*	*Rate of addition (kg ha^{-1})*
Zinc	200[a]/330[b]	250/420	300/500	450/750	15
Copper	80/130	100/170	135/225	200/300	7.5
Nickel	50/80	60/100	75/125	110/180	3
For pH 5.0 and above					
Cadmium	3/3(5)				0.15
Lead	300/300				15
Mercury	1/1.5				0.1
Chromium	400[c]/600[c]				15[c]
Molybdenum[d]	4/4				0.2
Selenium[d]	3/5				0.15
Arsenic[d]	50/50				0.7
Fluoride[d]	500/500				20

[a] Limit for samples taken to 25 cm (statutory) or 15 cm (monitoring).
[b] Limit for samples taken to 7.5 cm in grassland (monitoring).
[c] Provisional.
[d] Not statutory requirements.

Food chain contamination resulting from the use of sludge on grassland is likely to be minimal because heavy metals do not accumulate in either meat or dairy products for human consumption. The exception to this is cadmium which can accumulate in offal (liver and kidney). However, this is unlikely to be a problem where the EC soil limits are observed, especially since farm animals for meat production are short-lived. Certain lipophilic persistent organic contaminants, e.g. organochlorine insecticide residues (Lindsay, 1983), can accumulate in fatty tissues and be excreted in milk following direct ingestion of contaminated sludge. Suitable controls on trade effluents from industrial premises which make, use or handle these compounds usually restrict their levels in sludge to negligible concentrations. Therefore. specific limits for organics in relation to sludge utilization are not considered necessary at present. In the U.S.A., there is a limit of 10 ppm for PCBs in sludge used on land (US EPA, 1979), although the concentrations usually found are one to two orders of magnitude lower.

(b) Arable land

Sludge utilization on arable land is a well established outlet throughout the world. There is therefore a reliable base of operational experience and research results on which to develop environmentally acceptable practices suitable for particular agricultural conditions. In the U.K., arable land receives about 25% of sludge production. It is not as sensitive as grassland to problems of disease transmission, except when used on land growing crops eaten raw but problems can be avoided by appropriate sludge treatment and disposal practices (DoE, 1989).

The statutory limits for PTEs (Table 5.3) apply to soil samples taken to a depth of 25 cm, and these analyses, along with other information on sludge quality and recipient, must be kept on a register. The statutory soil samples must be taken before sludge is applied to land for the first time and then at extended intervals of 20 years since the accumulation of PTEs under normal circumstances is a very gradual process. The DoE Code of Practice also requires samples to be taken to 15 cm for regular monitoring purposes for which the same soil limits apply. The impact of PTEs on crops and the human diet is now well understood (see, for example, Carlton-Smith (1987)), although certain environmental questions remain to be resolved.

Leaching of nitrates presents a potential problem since there is increasing concern in Europe about groundwater quality (Commission of the European Communities, 1989) as well as in the U.K. over the proposed Nitrate Sensitive Areas. If implemented, this could seriously restrict sludge spreading on land in some areas. Thus it is important that sludge is applied in accordance with crop requirements for nutrients. Since sludge is produced throughout the year it is often convenient to spread it on land at times when the application of inorganic fertilizers would not be contemplated.

However, research indicates that sludge nitrogen being in the ammoniacal or organically bound form is not leached as readily as fertilizer nitrogen (Davis & L'Hermite, 1983). The potential problem of leaching of nitrate from sludge is put into perspective by the fact that about 1% only of agricultural land in Europe receives sludge annually. Thus concern should focus on the major nitrate souces, such as fertilizers and animal slurries. Guidelines recommend that sludge is not spread near aquifers or water abstraction points or where pollution of surface water could result.

The long-term effects of sludge in relation to soil quality and crop production also require examination. There has been concern that contaminants in sludge applied to the land may become more environmentally active and potentially toxic many years after applications have ceased. An increase in environmental activity might be linked to long-term changes in soil properties such as a reduction in organic matter or an increase in acidity. Evidence from one historically sludge-treated site suggests that metals from sludge may ultimately reduce soil microbial biomass and nitrogen-fixing capability (McGrath & Brookes, 1986). Research is in progress to find out if these effects are widespread and how they relate to crop production. The outcome of these investigations could lead to a further tightening of soil metal limits where sludge is used.

(c) Dedicated land

This is farmland set aside on a sacrifical basis for sludge disposal. The sites concerned are often old sewage farms surrounding the sewage treatment works, which have been used for the disposal of sewage and sludge over many years. Although these sites represent only a minor outlet for U.K. sludge they are of strategic importance locally and provide a very low-cost disposal outlet. Cultivation of forage crops of small-grain crops for animal feed is a low-risk option for historically contaminated soils in which metal levels may already exceed the maximum permissible concentrations. The EC Directive permits the continued use of dedicated land already in use in 1986 provided it is used to grow crops exclusively for animal consumption with no resulting hazard to either human health or the environment. Strict management along these lines (Rundle *et al.*, 1982) can permit productive farming to continue on sacrificial land which might otherwise be condemned to permanent dereliction. Apart from low cost, the advantage of using dedicated land is that any polluting potential is contained and contamination problems can be minimized by using suitable farming practices. The disadvantage is that the high concentrations of contaminants in the soil will severely restrict use of the land for other purposes. Drainage water from these sites will be high in nutrients, especially nitrate, following repeated heavy applications.

(d) Horticultural land and gardens

Use of sludge in gardens and for horticultural crops represents a comparatively high-risk outlet in terms of effects of metals and potential for pathogen
transmission because there is likely to be little control over how the sludge is
used. It will probably remain a minor outlet suitable only for sterilized or
composted sludge of very low metal content.

5.3.2 Land Reclamation

The DoE survey of derelict land for England and Wales indicates that there
are about 40 000 ha which justify reclamation largely as a result of mineral
extractions. Research in the U.K. and overseas (principally the U.S.A.) has
shown that the nutrients and organic matter supplied in sludge matches the
deficiences found in disturbed and derelict land (Byrom & Bradshaw, 1990;
Sopper, 1990), and thus not only improves the speed and quality of restoration but can be cheaper than conventional techniques (Metcalfe & Lavin,
1990).

The use of sludge can, for instance, avoid the need to bring in expensive
top soil since sludge can be mixed with soil-forming materials on site and the
resulting mixture used as top soil. Recent studies have focused on operational
aspects, such as methods of application and avoiding potential problems
(Hall *et al.*, 1986).

Currently, only about 5% of sludge is used in land reclamation, but
estimates suggest that the potential usage could be about 20% nationally,
although locally much higher since some areas have high concentrations of
dereliction (e.g. coal mining areas).

The principal constraints on expanding this outlet appear to be:

1. lack of appreciation of the value of sludge or derelict land by the
   restoration and water industries to their respective operations;
2. apprehension of potential environmental problems due to heavy metals,
   pathogens, odour, water pollution etc.;
3. logistical difficulties of matching continuous sludge production to irregu-
   lar and limited availability of restoration sites;
4. costs of disposal in relation to other local disposal options particularly if
   additional storage or special application equipment is required.

The environmental issues have been addressed by research and the
development of a code of practice (Hall, 1989), although further long-term
monitoring is desirable. Further research is required into the logistics of
supply and application since this not only affects the environment impact of
the operation, it also influences costs. A cost-benefit analysis is also required
to fairly distribute costs between supplier and recipient. The major need

in expanding the land reclamation outlet for sludge is educational and promotional.

5.3.3 Forestry

Since forests are planted generally on poor soils, tree growth is often limited by nutrients. Consequently the Forestry Commission apply about 30 000 t of fertilizer to 56 000 ha annually. Trials in the U.K. and U.S.A. have shown that sewage sludge, can be an effective alternative to conventional forest fertilizers (Bayes *et al.*, 1990, Nichols 1990). Ongoing research is studying growth response, environmental impact and operational aspects but experiences show that sludge can be utilized satisfactorily in commercial coniferous forests either to prepare the ground before planting (dewatered sludge) or to fertilize growing trees (liquid sludge). In these circumstances, there should be no problems due to pathogen transmission or food chain contamination. Environmental concerns relate to the possibility of run-off since forest land is often on steep slopes in water catchment areas, and to the acid soils which may cause metal mobility. According to the United States Environmental Protection Agency (1984) forest soils are well suited to sludge application, having high rates of infiltration (which reduce run-off and ponding), large amounts of organic material (which immobilize metals from the sludge), and perennial root systems (which allow year round application in mild climates).

Currently less than 1% of U.K. sludge is applied in forests although a number of new operations, principally in Scotland, have been set up recently. This may increase considerably, bearing in mind that currently 75% of sludge produced in Scotland is disposed of in the sea.

Forestry is expanding at around 30 000 ha per year, with a further 120 000 ha per year target for agricultural set-aside. There are also 12 new community forests planned to cover 40 000 ha.

Estimates by the Foresty Commission show that between 6 000 and 12 000 ha of existing forests would potentially be available annually assuming 16 and 32 km transport distance from sewage treatment works. Although most forests tend to be remote from urban areas, this could become an important outlet locally, and nationally could potentially account for up to 12% of annual sludge production. The potential of the proposed community forests has not been assessed but may be considerable since they will be located close to major conurbations, including coastal cities which currently dispose of sludge to sea.

5.3.4 Landfill

Disposal to sanitary landfills is used for more than 40% of sludge in Europe but only about 15% in the U.K. where landfill tends to be used on an *ad hoc* basis. In the U.K., most sludge is co-disposed with domestic refuse but in some countries, monofill is practiced. Co-disposal has received relatively little research attention and is not usually subject to specific disposal guidelines. The principal problem has been the poor physical nature of sludge resulting in handling and stability problems. This, and the reducing availability of landfill sites generally has tended to increase tipping charges.

There is a trend toward a minimum standard for physical properties in some European countries in order to avoid such problems (Van Den Berg *et al.*, 1990). However, in the U.K. attention has focused on improving handling techniques and identification of optimum ratios of addition to refuse (Hill 1990). Furthermore, there is increasing evidence that leachate quality may be improved by co-disposal (Farrell *et al.*, 1987) and that methane production is enhanced (Blakey, 1990). This is probably due to the buffering effect of sludge during the acid phase of anaerobic decomposition of the refuse allowing methanogenisis to occur many months earlier than without sludge. This effect will undoubtedly make sludge increasingly attractive for co-disposal where commercial recovery of methane as an energy source is envisaged.

5.3.5 Compost

Although, strictly speaking, composting is a treatment process rather than a disposal outlet, it alters radically the physical nature of sludge to such an extent that it opens up a wide range of sludge management and marketing options. Only small quantities of sludge are composted at present in the U.K. but it is used extensively in the U.S.A. Commercial interest is growing in the U.K. (see for example, Matthews & Border, 1990) particularly in response to public concerns over the depletion of our peatlands for production of plant growing media.

Composting is an aerobic treatment usually facilitated by mixing a bulking agent, often another waste such as straw, which in itself is likely to increase beyond its current surplus of 6–7 million tonnes per year following the banning of straw burning. The temperature during composting can rise to over 70°C so the process can achieve good control of pathogens but needs to be regulated by forced aeration or turning to achieve efficient and uniform treatment. The resultant compost is odourless, dry and friable compared with the clay-like consistency of dewatered sludge. Composting also reduces the bulk of material for disposal by about 50% so potentially reducing transport costs. Potential market areas identified are commercial horticulture, organic farming, tree planting and other amenity or land restoration purposes.

5.3.6 Minor and novel uses

There are a wide range of other uses for sludge which exploit its energy or chemical content. These have been reviewed by Frost & Bruce (1990) and Webber (1990). A major problem is the complex nature of sludge making the extraction or manufacture of useful substances difficult and prone to contamination.

Various types of sludge have been used as an animal feedstuff but with the possible exception of fish, this has not been successful. A range of potentially valuable constituents can be extracted from sludge such as protein, grease, vitamin B_{12}, metals and phosphorus but none is currently done commercially partly because of disposal problems of residues but largely because the economics are not attractive, being very susceptible to changes in world commodity prices. Production of fuel oil by low-temperature pyrolisis has been shown successfully, particularly in Canada (Campbell, 1989), and may become economically viable in the future when the price of energy increases.

The most successful minor outlets appear to be those that incorporate sludge into building materials. Sludge is incorporated into bricks ('Biobrick') and fibre board, and sludge incinerator ash is used as lightweight aggregate ('Sludgelite') and filler for clay pipes and tiles (Murakami & Oshima, 1988). None of these outlets is likely to be more than of local importance at least in the short-term, and all require an 'entrepreneurial spirit' on the part of the sludge producer to promote an unusual outlet for his product.

5.3.7 Incineration

There have been considerable improvements to the technology of incineration, particularly in Europe, and modern fluidized bed incinerators now appear more attractive in terms of both capital and operating costs compared with the multiple hearth type. Techniques are now available to control gaseous emissions to meet stringent environmental standards. Until recently only about 4% of U.K. sludge was incinerated by four incinerators, but two new plants have been constructed in Yorkshire (Lowe & Boutwood, 1990), and many water authorities are examining the option in the light of the phasing out of sludge disposal at sea.

Incineration has previously been seen as the last option, for instance, where sludge has been too heavily contaminated to go to land or where road haulage would cause nuisance. With the development of efficient autothermic incineration requiring no other fuel source except for start up, incineration costs are becoming much more competitive with other disposal options, to the extent that incineration is now seen by some as the only solution to the increasingly intractable problems of other sludge disposal options. There is also interest in co-incineration with other waste materials, in power stations and in cement production, but it is unlikely that there will be much take up of such options at least in the near future.

However, incineration does not provide complete disposal since about 30% of the solids remain as ash. This is generally landfilled and is regarded in some countries as a highly toxic waste because of its high metal content. The major constraint on more widespread use of incineration will be the planning difficulties associated with public concern about possible emissions.

5.3.8 Sea disposal

Currently, sludge is disposed of by ship to disposal grounds licensed and monitored by the Ministry of Agriculture, Fisheries and Food, under the Food and Environment Protection Act (1985). The overall contribution by sludge to the pollutant load entering the sea is low and consequently environmental effects have proved difficult to evaluate. The U.K. has now agreed to stop sea disposal by the end of of 1998 following growing political pressure from other EC member states. It is also increasingly likely that treatment of sewage discharge through sea outfalls will be required in the future. If so, this will not only increase costs but also produce even more sludge for disposal on land.

5.4 CONCLUSIONS

The need to dispose of sludge economically and safely will be a continuing requirement into the future. The challenge facing the sludge disposal authorities is to find a cost-effective compromise between the conflicting requirements of reducing costs and increasing value for money on the one hand, while responding to the effects of increasing environmental pressures on the other. An innovative technological response will be needed.

It is possible to envisage, as a consequence of phasing out the disposal of sludge at sea, that much greater quantities of sludge will be incinerated and landfilled; options used more widely in Europe and the U.S.A., although it is now US EPA policy to recycle sludge wherever this is feasible. Incineration will require a long lead-in time even where planning permission is forthcoming, and landfilling will expand in the short-term but will be limited by site availability. Fig. 5.3 shows how these possible changes to the pattern of sludge disposal may take place over the next 20 years. Outlets such as land reclamation and forestry will expand considerably locally but will probably remain relatively small outlets nationally unless the development of incineration is restricted due to public pressure.

The possibilities for the re-use of wastes, such as sewage sludge, will become increasingly difficult in the future because of the inevitable presence of contaminants and ever more stringent environmental standards, unless

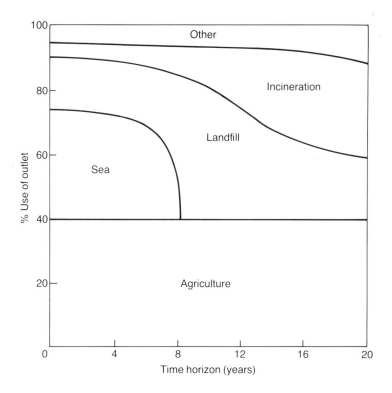

Figure 5.3 Possible future changes to the pattern of sludge disposal.

methods of avoiding or removing contaminants can be developed. The prospects of suitable treatment processes to remove contaminants, such as heavy metals, appear remote at present, but equally difficult will be the reduction of such inputs to the sewer except for specific industrial sources. Most of the contaminants in sludge reflect very many diffuse and low level sources which will prove impossible to control or remove without widespread and radical alterations to the composition of domestic products, and other environmental inputs, such as atmospheric deposition, which find their way into the sewerage system. Furthermore, many of the valuable components in sludge, such as nutrients, are also pollutants if they are transferred from soil to the water environment. Thus even if low metal sludge could be attained in the future, its re-use would still have to be strictly regulated so as to avoid other undesirable effects to the environment.

The selection of the best practicable environmental option involves consideration of social as well as environmental, economic and technical factors. The general public is becoming increasingly well-informed on environmental matters and environmental quality is a political issue. While keeping within economic constraints, operators will have to ensure that sludge is disposed of or recycled in an environmentally acceptable way and that the public perceives that this is so.

REFERENCES

Bayes, C.D., Taylor, C.M.A. and Moffat, A.J. (1990) Sewage sludge utilisation in forestry: the UK research programme. In *Alternative uses for sewage sludge* (ed. J.E. Hall). Pergamon Press.
de Bekker, P.H. and van den Berg, J.J. (1987) Landfilling with sewage sludge. In *Treatment of sewage sludge: thermophilic aerobic digestion and processing requirements for landfilling* (ed. A.M. Bruce, F. Colin and P.J. Newman), pp. 72–79. London: Elsevier Applied Science.
Blakey, N.C. (1990) Co-disposal of sewage sludge and domestic waste in landfills: laboratory and field trails. In *Alternative uses for sewage sludge*. (ed. J.E. Hall), Pergamon Press.
Bowden, A.V. (1987) Survey of European sludge treatment and disposal practices. *Water Research Centre Report 1656-M*. Medmenham: Water Research Centre.
Bruce, A.M., Campbell, H.W. and Balmer, P. (1984) Developments and trends in sludge processing techniques. In *Processing and use of sewage sludge* (ed. P. L'Hermite and H. Ott), pp. 19–38. Dordrecht: D. Reidel.
Byrom, K. and Bradshaw, A.D. (1990) The potential value of sewage sludge in land reclamation. In *Alternative uses for sewage sludge* (ed. J.E. Hall). Pergamon Press. (In the press.)
Campbell, H.W. (1989) A status report on environment Canada's oil from sludge technology. In *Sewage sludge treatment and use: new developments, technological aspects and environmental effects* (ed. A.H. Dirkzwager and P. L'Hermite), pp. 281–290. London: Elsevier Applied Science.
Carlton-Smith, C.H. (1987) Effects of metals in sludge-treated soils on crops. *Water Research Centre Technical Report 251*. Medmenham: WRc.
Commission of the European Communities (1983) *Concerted Action: treatment and use of sewage sludge (COST 68 ter). Final report of the community – COST concertation committee*. I Scientific Report. Brussels: CEC.
Commission of the European Communities (1988) *Proposal for a council directive concerning the protection of fresh, coastal and marine waters against pollution caused by nitrates from diffuse sources*. COM (88) 708.
Commission of the European Communities (1989) *Proposal for a council directive concerning municipal wastewater treatment*. COM (89) 518.
Council of the European Communities (1986) *Council directive on the protection of the environment and in particular of the soil, when sewage sludge is used in agriculture. Off. J. Europ. Comm.* L **181**, 6–12.
Davis, R.D. and L'Hermite, P. (eds). (1984) *Utilisation of sewage sludge on land: rates of application and long-term effects of metals*. Dordrecht: D. Reidel.
Davis, R.D., Carlton-Smith, C.H., Stark, J.H. and Campbell J.A. (1988) Distribution of metals in grassland soils following surface applications of sewage sludge. *Environ Pollut*. **49**, 99–115.
Department of the Environment (1989) *Code of practice for agricultural use of sewage sludge*. London: H.M.S.O.
Department of the Environment/National Water Council (1981) Report of the sub-committee on the disposal of sewage sludge to land. *DoE/NWC Standing Tech. Comm. Rep*. **20**. London: DoE.

Farrell, J.B., Dotson, G.K., Stamm, J.W. and Walsh, J.J. (1987) The effects of municipal wastewater sludge on leachates and gas production from sludge-refuse landfills. Presented at the US/USSR bilateral agreement symposium on municipal and industrial wastewater treatment, March 20–21. Breidenbach, Environmental Research Center, Cincinnati, Ohio, U.S.A. Cincinnati: USEPA.

Fisher, W.J. (1971) Oxidation of sewage with air at elevated temperatures. *Wat. Res.* **5**, 187–207.

Frost, R.C. and Bruce, A.M. (1990) Energy from sludge. In *Alternative uses for sewage sludge* (ed. J.E. Hall) Pergamon Press.

Hall, J.E. (1988) Application of sewage sludge to agricultural land. A directory of equipment (2nd edn). Water Research Centre Report PRU 1857. Medmenham: Water Research Centre.

Hall, J.E. (1989) Methods of applying sewage sludge to land: a review of recent developments. In *Sewage sludge treatment and use: new developments, technological aspects and environmental effects* (ed. A.H. Dirkzwager and P. L'Hermite), pp. 65–84. London: Elsevier Applied Science.

Hall, J.E., Daw, A.P. and Bayes, C.D. (1986) The use of sewage sludge in land reclamation. A review of experience and assessment of potential. Water Research Centre Report 1346–M. Medmenham: Water Research Centre.

Hibberd, S. (1863) *Profitable gardening*. London.

Hill, C.P. (1990) The co-disposal of controlled wastes and sewage sludge – some practical aspects. In *Alternative uses for sewage sludge* (ed. J.E. Hall). Pergamon Press.

Holladay, D. (1986) Review of sewage sludge screening in the U.K. *Water Research Centre Report 470S*. Medmenham: Water Research Centre.

Hoyland, G. and Ronald, D. (1984) Biological filtration of finely-screened sewage. *Water Research Centre Report TR 198*. Medmenham: Water Research Centre.

Lichensteiger, T., Bruner, P.H. and Langmeier, M. (1987) Transformation of sewage sludge in sanitary landfills. In *Treatment of sewage sludge; thermophilic aerobic digestion and processing requirements for landfilling.* (ed. A.M. Bruce, F. Colin and P.J. Newman), pp. 58–71. London: Elsevier Applied Science.

Lindsay, D.G. (1983) Effects arising from the presence of persistent organic compounds in sludge. In *Environmental effects of organic and inorganic contaminants in sewage sludge.* (ed. R.D. Davis, G. Hucker and P. L'Hermite), pp. 19–26. Dordrecht: D. Reidel.

Lowe, P. and Boutwood, J. (1990) The use of sewage sludge as a fuel for its own disposal. In *Alternative uses for sewage sludge.* (ed. J.E. Hall). Pergamon Press.

McGrath, S.P. and Brookes, P.C. (1986) Effects of long-term sludge additions on microbial biomass and microbial processes in soil. In *Factors influencing sludge utilisation practices in Europe.* (ed. R.D. Davis, H. Haeni and P. L'Hermite), pp. 80–89. London: Elsevier Applied Science.

Matthews, P.J. and Border, D.J. (1990) Compost – a sewage sludge resource for the future. In *Alternative uses for sewage sludge.* (ed. J.E. Hall). Pergamon Press.

Mechi, J.J. (1859) *How to farm profitably*. Essex: Tiptree.

Metcalfe, B. and Lavin, J.C. (1990) Consolidated sewage sludge as soil substitute in colliery spoil reclamation. In *Alternative uses for sewage sludge.* (ed. J.E. Hall) Pergamon Press.

Murakami, K. and Oshima, Y. (1988) Treatment, disposal and utilisation of sewage sludge in Japan – current practice and future direction. Proceedings 3rd WPCF/ JSWA Joint Technical Seminar on Sewage Treatment Technology, Tokyo.

Nichols, C.G. (1990) US forestry uses of municipal sewage sludge. In *Alternative uses for sewage sludge.* (ed. J.E. Hall). Pergamon Press.

Poulanne, J. (1984) Sludge production rates. In *Processing and use of sewage sludge.* (ed. P. L'Hermite and H. Ott), pp. 39–50. Dordrecht: D. Reidel.

Rundle, H, Calcroft, M. and Holt, C. (1982) Agricultural disposal of sludges on a historic sludge disposal site. *Wat. Pollut. Cont.* **81**, 691–632.

Sopper, W.E. (1990) Utilisation of sewage sludge in the United States for mine land reclamation. In *Alternative uses for sewage sludge.* (ed. J.E. Hall). Pergamon Press.

Statutory Instruments (1989) *The sludge (use in agriculture) regulations.* **1263.**

United States Environmental Protection Agency (1979) *Fed. Regist.* **44**, 53455.

United States Environmental Protection Agency (1984) *Use and disposal of municipal wastewater sludge.* Cincinnati: USEPA Office of Research Management.

United States Environmental Protection Agency (1989) *Fed. Regist.* **54**, 5746–5902.

Van Den Berg, J.J., Geuzens, P. and Otte-Witte, R. (1990) Physical aspects of landfilling sewage sludge. In *Alternative uses for sewage sludge* (ed. J.E. Hall). Pergamon Press.

Water Research Centre (1985) *The agricultural value of sewage sludge. A farmers' guide.* Medmenham: Water Research Centre.

Water Research Centre (1989) *Soil injection of sewage sludge. A manual of good practice* (2nd edn). FR 0008. Medmenham: Water Research Centre.

Webber, M.D. (1990) Resource recovery through unconventional uses for sludge. In *Alternative uses for sewage sludge.* (ed. J.E. Hall). Pergamon Press.

6

Municipal solid waste management: an economic perspective

R.K. Turner

6.1 INTRODUCTION

The late 1980s have witnessed a reorientation of official environmental protection policies, paralleling a deepening sense of public concern over the future prospects for a sustainable economic and environmental system. The Treaty of Rome, as modified by the Single European Act, places environmental policy among the official policies of the European Community (EC). Article 130r(2) lays down, among other things, that environmental protection requirements shall henceforth be a component of the Community's other policies, with regard to agriculture, trade, industrial development, etc. The needs of the environment therefore constitute an important facet of the internal market. Waste management activities (waste minimization, recycling, transport and final disposal) will have an increasingly important impact on the functioning of the internal market. The EC's Fourth Action Programme (1987–1992) confirmed three integrated lines of policy:

1. waste prevention;
2. waste recycling and re-use;
3. safe disposal of non-recoverable residuals.

In the U.K., recycling has been given a prominent position in the government's new blueprint for environmental protection in the 1990s and beyond. The Environmental Protection Act 1990 has been designed to replace Part I of the 1974 Control of Pollution Act. The 1990 Act seeks to clarify,

The Treatment and Handling of Wastes
Edited by A.D. Bradshaw, Sir Richard Southwood and Sir Frederick Warner
Published in 1992 by Chapman & Hall, London, for The Royal Society
UK ISBN 0 412 39390 5, USA ISBN 0 442 31461 2

strengthen and extend official controls over waste. It places a duty of care on producers and handlers of waste in the private and public sectors. Waste collection and regulation authorities are to be given specific duties in the context of the recycling of municipal wastes in their respective areas. A recycling statement or plan will be required setting out the arrangements made or proposed. It will be the duty of an authority to have regard to the desirability, where reasonably practicable, of giving priority to recycling waste.

The main purpose of this paper is to examine, from an economic perspective, the issues surrounding the choice of the most appropriate future system of municipal solid waste management, given that there are a number of alternative configurations all of which are technically feasible (see Fig. 6.1).

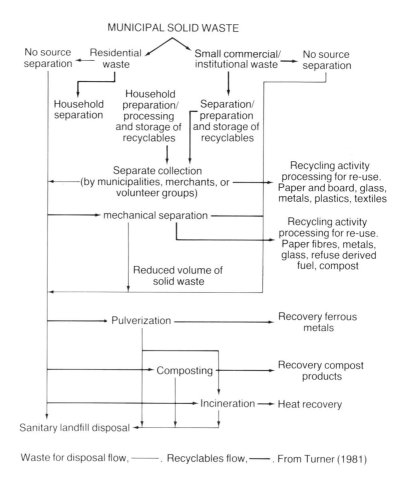

Waste for disposal flow, ———. Recyclables flow, ——. From Turner (1981)

Figure 6.1 Simplified municipal waste management.

Currently, the management system is still mainly based on the conventional 'dispose, dilute and disperse' approach (centred on landfill disposal). All the indications are that a fundamental philosophical switch is required, towards a 'recycle, concentrate and contain' approach. It will be argued in succeeding sections of this paper that economic analysis (and the emerging principles of so-called sustainable economic development) supports such a reorientation in waste management policy. Landfill disposal will remain an important component of any future waste management strategy. However, the landfill of the future will be a much more sophisticated and 'engineered' set of operations than it has been in the past. Consequently, the average costs of waste disposal via landfill are set to rise sharply. This rise, reflecting the true social costs of waste disposal, will have important positive ramifications for waste minimization and waste recycling schemes.

6.2 GENERAL OBSERVATIONS ON THE EVALUATION OF OPTIONS

Currently, the statutory duty of a waste disposal authority is to ensure that satisfactory arrangements are made available for the safe and hygienic disposal of controlled waste for which it has responsibility. In general terms the overall objective of a waste management strategy would be the disposal of waste at the least possible cost to the community, with due regard to the safeguarding of the environment and the use of waste as a resource. But translating this broad statement into precisely defined criteria against which alternative technical options and disposal plans can be assessed is not straightforward.

This is also true of the Best Available Technology Not Entailing Excessive Costs (BATNEEC) principle which governs the licensing of new waste disposal facilities. Here there is a clear implication that economic and environmental objectives somehow have to be reconciled. Identifying the Best Practicable Environmental Opton (BPEO) for elements of the waste management system, or more ambitiously the system itself, is the ultimate long-term challenge for a sustainable economic system (Royal Commission on Environmental Pollution, 1988; Department of the Environment, 1988). Complications have arisen because of three main factors:

1. these days it is no longer acceptable to make decisions on solid waste management merely by reference to financial costs and benefits within a limited and prescribed set of physical constraints such as available landfill disposal capacity. A range of social and environmental aspects must also be considered;
2. the growing scale and technical complexity of potential waste treatment systems;
3. the increasing range of technical solutions.

Rational decision-making has also been made more difficult in the U.K. because of information failure. Accurate projections for the quantities and composition of solid waste are essential for the planning of efficient and economical solid waste collection, processing and disposal systems. The main source of official data (discontinued in 1986–87) on waste collection and disposal in England and Wales, the Cipfa statistics (Chartered Institute of Public Finance and Accountancy: Statistical Information Services, annual returns) suffered from a number of deficiencies, despite the increased percentage of the returns over time said to be based on actual weight data. More accurate total waste generation and waste composition data are urgently required.

The options and plans can be assessed in terms of financial cost-effectiveness. Recycling schemes would be evaluated according to their financial profitability, and disposal options according to their net financial cost per tonne of waste. Schemes generating the highest net profitability and the least costly disposal options would have high priority in the management strategy. Alternatively, waste management could be subjected to an economic evaluation via the medium of cost-benefit analysis.

6.3 MARKET FORCES AND WASTE MANAGEMENT

A significant proportion of the municipal waste stream which could potentially be recovered (recycled) is in fact discarded, because under current market conditions the financial costs of recovery exceed the financial value of the recoverable materials and/or energy. It is important to understand that the U.K. and other industrialized economies have traditionally been rather good at recycling some types of secondary (scrap) materials. The combined activities of long-established private waste reclamation industries and a wide variety of municipal and charity group recycling ventures have served to push national aggregate materials recycling rates up to fairly high levels throughout Europe and North America.

Intuitively, it seems reasonable to suppose that a currently high overall recycling rate shows that the least-cost recovery options have been exploited and that further recycling stimulated by government intervention may prove financially costly. Research work has shown that this generalization is essentially valid in many specific cases (Turner, 1983).

The relatively high national recycling effort is largely because of the activities of the private reclamation industry, which concentrates its attention on industrial and commercial waste flows. Typically, these wastes will be generated in large concentrated quantities, they will be of known composition and will be relatively uncontaminated. The favourable nature of these physical characteristics (mass, contamination, homogeneity and location) combine to increase the prospects of a financial profit for the would-be reclaimer.

On the other hand, the physical characteristics of the various components of municipal waste streams are such as to make recycling less financially attractive. The potential resources in municipal waste have not been heavily recycled because this source of secondary materials yields more dissipated and usually more contaminated materials. Thus, more often than not, financial costs of recovery (in particular costs of collection) exceed the financial value of the recovered materials and/or energy.

The U.K.'s recovery rate for components of the municipal solid waste stream is low. Estimates vary from around 2% of the total waste stream to between 5–10% if civic amenity sites are included. In 1987 the U.K. consumption of *waste paper and board* reached a record level of 2.417 million tonnes. This represented a recovery rate of about 30%, which is only slightly above the 28% rates that were achieved as far back as the 1970s. The U.K. has one of the lowest and slowest growing recovery rates in Europe, which supports one of the highest utilization rates. The U.K. has a high consumption of paper and board per head, but imports 60% of it. Surplus waste paper is exported.

Glass (cullet) recycling also reached a new record level in 1988, at some 275 000 tonnes. This represented a recovery rate of 16% which again was well below European rates. The Netherlands and Switzerland, for example, have recovery rates in excess of 50% and many more bottle banks per head of population.

The total quantities of aluminium (foil and cans) lost to the municipal waste-stream is around 150 000 tonnes per annum. Only about 1% of cans are currently being recovered, whereas in Germany the recovery rate is 33%.

Plastics recycling has been inhibited by technical constraints (only thermo-plastics such as PET bottles can be recycled). Around 820 million PET bottles were sold in 1987 as opposed to only 128 million in 1980, and PET now accounts for 34% of all the soft drinks packages sold, yet recycling of such items is on a very small scale at present.

It has been estimated that some 30 000 tonnes of CFCs are locked up in U.K. refrigeration and air conditioning systems. Some 10% of this total is accounted for by domestic fridges and freezers. Schemes for equipment collection and recycling of CFCs have yet to be established on any significant scale. Although recovery of CFCs contained in refrigerant is possible, only 5% or less of redundant units are being drained. It is estimated that some 1500 tonnes of CFCs are being released into the atmosphere annually in the U.K. via this route.

Theoretically, in the absence of government intervention in the form of an environmental quality management policy, materials or energy recovery will take place up to the level where the marginal cost of an additional unit of recycled material or energy just equals its market value in a reclaimed condition. Both the recovery costs (collection plus processing costs) and the value of the recovered item will be subject to change over time. In reality, the picture becomes more complicated because the government does intervene in the market mechanism for a number of reasons, including environmental quality protection. The extent to which materials and energy recycling is practised will be a function of the cost of recovered residuals as a raw material input into production, relative to primary raw material inputs. These relative costs will be determined by a complex flux of factors including: technological innovation in primary and secondary process industries; technical advances in primary material extraction; secondary and primary material market conditions and structure; end-product output specifications; and government policy on environmental quality protection and in particular on final disposal of wastes.

6.4 THE EXAMPLE OF WASTE PAPER AND BOARD

The current (1989–90) difficulties surrounding paper and board production recycling activities are illustrative of the complexities involved. Since the early 1970s the real cost of waste paper has fallen and the cost of primary pulp from Canada, for example, has risen threefold. In 1987–88 it was forecast that the U.K. would be facing a shortfall of some 700 000 tonnes of waste paper and board by 1993. The forecast was based on the assumption that new technology (advances in de-inking technology) and extra paper-making capacity were coming on stream. Both the proportion of waste paper per unit of final output and the number of grades of waste paper suitable as feedstock were expected to increase.

As it has turned out, the first half of 1989 saw a growing over-supply problem, at least as far as lower grade waste paper and board (news and pams) are concerned. The international waste paper market is also in excess supply as North American exports of lower grade paper (generated by mandatory municipal source separaton schemes) have increased significantly. U.K. exports of waste paper have consequently been inhibited and domestic prices for used newsprint have collapsed.

Technological change brings both costs and benefits. The new de-inking technology has allowed the proportion of lower grade waste paper in products such as tissues and paper towels to increase. But U.K. manufacturers have not yet made the decision (on production or marketing grounds) to produce some types of tissues from de-inked newsprint, as Swedish plants do. Other technological advances such as flexographic and laser printing, latex adhesive and the continued use of clay filters in paper all serve to inhibit recycling. Furthermore, effluent from the recycled fibre operations poses a threat to water quality, both in terms of its BoD demand and production of dioxins.

Unless end-product quality standards such as paper 'brightness' are changed and product design encompasses recyclability, the supply of higher grade paper waste will continue to limit the range of end-products made from substantial inputs of recycled inputs. By 1993 the situation, however, will have partially changed as far as newsprint is concerned. Newsprint mills will by then be in a position to meet 50% (as opposed to the current 30%) of home demand and will require around 1 million tonnes of lower grade waste paper feedstock. It remains an open question whether there are substantial untapped supplies of higher grade commercial or office papers in the U.K.

6.5 RECYCLING VERSUS DISPOSAL: THE PRIVATE COST (FINANCIAL) APPRAISAL

Local authorities inherit a waste-stream of municipal solid waste from local waste generators (households, shops, offices, schools, etc.) and have a statutory obligation to collect or dispose of the waste. Let us now take a closer look at the issue of whether or not components of this should be recycled rather than disposed. Assuming that it is technically possible to recycle a certain proportion of a range of materials from municipal waste (which it certainly is for paper, glass, some plastics and metals) the first question that needs to be asked is, is there a market for these reclaimed materials? Again, the answer is in the affirmative for glass, metals and paper, and on a more limited basis for plastics. Because there is a market, processors will buy in the waste material at some price determined by market conditions. Recycling material will under these conditions generate a financial revenue. But the recycling process itself (materials separation, collection and transport) will also use up resources and therefore will involve financial costs (wages, fuel, vehicles, etc.). It is possible then to calculate the net financial cost (positive or negative) for the recycling scheme.

If the waste (or some components of it) had not been recycled it would have had to have been disposed of to landfill or incineration. If the disposal authority is rational it will seek the least cost (to it) disposal option. Recycling makes financial sense if the total net financial costs of the recycling and disposal option is less than the financial costs of the least cost disposal option. This is more likely to be the case the higher is the cost of disposal incurred by the waste disposal authority. Thus for metropolitan communities lacking suitable landfill disposal sites in close proximity to waste generation sources, disposal costs will be relatively high. They have to pay for the bulk haulage of waste over long distances, which is relatively expensive.

6.6 RECYCLING VERSUS DISPOSAL: THE SOCIAL COST (ECONOMIC) APPRAISAL

Standard economic analysis is based on an efficiency criterion which suggests that because recycling activities are not themselves costless, technically feasible recycling should take place up to the point at which total social benefits exceed total social costs by a maximum amount. Economic efficiency therefore has to do with obtaining the greatest value of real output for a given input cost, or alternatively achieving a given value of real output at minimum input cost. The concept of a social cost or benefit encompasses both the financial costs of, or revenues derived from, a recycling scheme and any environmental cost or benefit. Quantified social costs and benefits should then all be placed on a common monetary scale of value.

For the market's efficiency criterion to be satisfied and in line with the Polluter Pays Principle (Pezzey, 1988), all the social costs of production and consumption must be reflected in the producer's costs of production and in the market price paid by consumers. Real world markets often fail to transfer the full social costs of production and consumption to the producers and consumers. In particular, the waste assimilation services provided by the environmental media (land, air and water) are sometimes treated as almost free goods and are therefore overexploited.

Thus if landfill disposal in the future is more stringently regulated, to minimize long run pollution risks, the average (per tonne) cost of disposal will rise significantly. The introduction of landfill gas venting and monitoring, steel sheet piling to minimize groundwater pollution and a more precise costing of waste transport costs, increases average costs from £3–4 per tonne to at least £16 (1988–89 prices). Other disposal options, such as energy-from-waste schemes begin to look competitive at this cost level.

Waste disposal authorities need not be required to prove the financial profitability of any given recycling scheme, but merely that the introduction of such a scheme into the waste management system has reduced the overall net social costs of the system. The authorities have a statutory duty to collect and dispose of municipal waste, and so it is the relative costs of recycling compared with waste collection and disposal that is significant.

To be more precise, a recycling scheme need only compare well with the currently available least-cost disposal alternative in practice. In the U.K. this will be landfill disposal. A recycling scheme will generate overall net social benefits, as long as the net social costs of the least-cost disposal option (landfill) are greater than the net social costs associated with the combined recycling and disposal option. Thus, waste disposal authorities which are already facing, or are likely to be facing, high disposal costs (probably involving long distance bulk haulage) will, *a priori*, be the best candidates for combined recycling/disposal options.

Research work carried out in a number of different countries has indicated that recycling can yield the following potential benefits (in terms of the avoided costs of the alternative option i.e. the least-cost disposal option):

1. savings in refuse collection and/or disposal costs, depending on the form of organization and type of disposal option adopted (disposal cost savings will include reduced level of pollution and disamenity costs);
2. reduction in most cases in overall pollution impact, when recycled materials, are used as raw material inputs in production processes;

3. reduction in the quantity of primary materials requiring extraction and processing, as well as consequent reductions in energy use (Turner, 1981).

However, the recycling activity itself is not a costless exercise. Whatever the actual recycling system chosen, there will be some combination of extra material collection costs and/or processing and transportation costs. Potential pollution costs relating to recycling must also be taken into account.

We can now itemize, in a more formal way, both the benefits of recycling (in terms of the avoided costs of the disposal option) and the costs of recycling.

$$\text{Benefits of recycling } = BR = V_C + V_P + V_F + S_K + S_D, \qquad (6.1)$$
$$\text{or } BR = SCP + SCD$$

where V_C = private costs of primary material extraction (extractions costs + user costs);

$V_P + V_F$ = pollution/environmental costs of primary material extraction/harvest and processing;

$S_K + S_D$ = cost of collection for disposal purposes, and the cost of disposal (including pollution and disamenity costs S_{DP});

SCP = total social costs of primary material use;

SCD = total social cost of collection and disposal system without the recycling option.

$$\text{Cost of recycling} = SCS = S_{Kr} + S_{St} + S_P, \qquad (6.2)$$

where S_{Kr} = cost of collection for recycling purposes;

S_{St} = cost of separation, processing and transportation of recycled material;

S_P = pollution costs of recycling,

and SCS = total social costs of recycling activity.

Hence, net benefits of recycling, NBR, are given by:

$$NBR = V_C + V_P + V_F + S_K + S_D - S_{Kr} - S_{St} - S_P \qquad (6.3)$$
$$\text{or } NBR = SCV + SCD - SCS.$$

This general formula now needs to be adapted to take into account:

1. different alternative physical recycling system configurations (i.e. source separation scheme variants or centralised, mechanical separation plant variants);
2. different recyclable materials, paper, glass, aluminium, tinplate/steel, and plastic.

Equation (6.3) can be simplified and rearranged to give us:

$$NBR = (V_C - S_{St}) + (V_P + V_F + S_{DP} - S_P) + (S_K - S_{Kr}) + S_D \qquad (6.4)$$

$V_C - S_{St}$ is now the private cost component reflecting the difference between the cost of primary materials and the cost of delivered secondary materials net of collection costs;

$V_P + V_F + S_{DP} - S_P$ is the net environmental impact, where S_{DP} is the pollution/disamenity costs of disposal;

$S_K - S_{Kr}$ is the net collection cost, given the adoption of a recycling scheme;

and S_D, is the possible savings in disposal costs credited to the recycling scheme (excluding pollution/disamenity costs associated with disposal).

6.7 PHYSICAL RECYCLING SCHEME OPTIONS: SOURCE SEPARATION VERSUS MECHANICAL SEPARATION SCHEMES

In terms of the private costs ($V_C - S_{St}$), V_C is the same whatever physical configuration is chosen to enable recycling to take place. But S_{St} is likely to be lower if source separation, rather than mechanical separation options are adopted. This is because the mechanical plant has to deal with a mixed waste-stream (higher relative entropy), before separation and processing can take place. By definition, source separation keeps materials separate from each other from the start. There is of course, a hidden cost in source separation schemes, in terms of extra time/inconvenience and storage space costs internalized by individual households participating in the scheme. On the other hand, these costs may be cancelled out by the 'benefit', participating households may feel in terms of making a contribution to an environmentally acceptable policy.

In source separation schemes, where 'drop-off' centres are being used, separation and processing costs (S_{St}) can be further reduced by the use of voluntary labour, handicapped labour or long-term unemployed labour.

In terms of the net environmental impact ($V_P + V_F + S_{DP} - S_P$), $V_P + V_F$ is the same for source separation and mechanical separation systems. S_{DP} will, however, be potentially higher if the mechanical separation option is chosen, and proves to be technically feasible/reliable. This is because the residual unrecycled waste tonnage left over and requiring disposal, is much lower under the mechanical separation option. Thus, there are less potential pollution/disamenity costs.

As far as S_P is concerned, it is not easy to compare pollution impacts on the basis of recycling differences. There is, however, some concern about possible air pollution impacts from mechanical recycling plants if this category is taken to include energy from waste systems.

In terms of the net collection costs ($S_K - S_{Kr}$), provided that the mechanical separation plant is centrally located there should be little if any change in normal (disposal) system collection costs. An exception would be a situation where a centralized plant required a large geographical catchment area, because of the need to operate on an economic waste throughput volume, and was faced with dispersed waste generation sources.

Some variants of the source separation option (ie. door-to-door collections, both integrated with or separate from normal refuse collection) will result in $S_{Kr} > S_K$, that is, a net increase in collection costs. If a drop-off or some other central container-type collection system is used, then, in principle, $S_K - S_{Kr}$ will represent a net saving in collection costs. In practice, such savings may be negligible, because of indivisibilities of collection in the short or medium term. In schemes where a net increase in collection costs is incurred, these increases can be minimised by the use of voluntary labour.

Disposal cost savings (S_D), will be greater under the mechanical separation option provided that it proves technically feasible to remove, glass, ferrous and non-ferrous metals, utilize all the light fraction (paper/plastics) as an energy source, and compost some putrescible matter, because of the small residual waste tonnage requiring disposal. So far, no plant has been able to operate on this basis consistently over a period of time, and so S_D will not differ significantly under different physical recycling system configurations.

Some caveats are in order before the general recycling evaluation formula is applied to specific materials found in municipal solid wastes. Secondary material inputs are often not perfect substitutes for their primary material counterparts. Thus, for example, one tonne of lower-grade waste paper is not equivalent to one tonne of mechanical pulp (there is 20% fibre loss involved). Similarly, recycled aluminium inputs cannot substitute, across-the-board, for primary aluminium, and some plastics cannot be recycled at all, except as an energy source. The existence of a grade structure in recycled materials tends to constrain the marketability of recycled materials because of limited end-uses, and therefore lowers their financial value (e.g. paper or board and plastics). The output from mechanical separation plants (low grade mixed paper, glass and compost) has proved difficult to market.

On the other hand, some recycled materials (e.g. aluminium, glass and to a lesser extent paper or board) yield a benefit in terms of lower final product manufacturing process energy requirements ($V_{MC} > S_{MC}$), and $V_{MC} - S_{MC}$ = process energy savings credited to recycled inputs.

6.8 RECYCLABLE MATERIALS IN MUNICIPAL SOLID WASTE

(a) Paper and Board

Does paper recycling save trees and therefore V_C? The answer depends on current and future trends in global afforestation. If it is the case that the harvesting of 'natural' forests has, or is about to, reach a level where the harvest rate is in excess of the sustainable yield (i.e. forest stocks are being reduced) then recycling can contribute to forest preservation. In this context there will be benefits in terms of V_C and V_F. This conclusion is reinforced by evidence indicating that natural forests in developing countries are now being exploited as new sources of pulp. The greenhouse effect problems serve to increase the forest preservation benefits of recycling, and are an element of V_P.

The U.K. imports significant quantities of pulp from Scandinavia. These forests are being managed on a sustainable basis and therefore the user cost element in V_C will be zero. The position in some developing countries (with hardwood forests) is different and in these cases, where hardwood forests are being used for pulp, operations are unsustainable. Hence, user cost will be positive in this context. Whether, in practice, recycling will save tropical forests is open to debate. The logging of such forests is only one cause of deforestation (other probably more important factors include land clearance for agriculture). It is also possible to argue that in some circumstances recycling could reduce greenhouse credit by reducing incentives for afforestation.

Trying to assess the net pollution impact, $V_P - S_P$, is more difficult. If the whole process of paper manufacture from timber felling to final product is considered, there may well be pollution savings because of recycling. However, where an economy, such as the U.K., imports woodpulp, it may find that large elements of the environmental impact have occurred in the pulp manufacturing country. In this respect, a comparison of the use of woodpulp and recycled fibre is less obviously in favour of recycling in the importing country. Much will depend on the stringency of water pollution controls covering de-inking plant effluent discharges (BOD, toxic dioxins and adsorbable organic halogens).

It is not possible to state unambiguously that waste paper recycling has energy savings benefits. Again, much depends on whether one is taking a national or international standpoint. For the U.K. (given its pulp deficit situation) an important comparison is between imported pulp and domestic waste paper inputs into production. In this context, if there are energy savings they are probably not large, as the energy intensity is partly taken up in the pulp producing country. If the entire process from timber felling to paper manufacture is considered, it seems probable that recycling will save energy.

A further complication arises because relative energy cost advantages vary when individual pulp-based products are compared with waste-paper based products. Thus in newsprint production, there is a clear energy saving associated with the use of waste paper, but for other products the balance of advantage is not as clear cut.

(b) Glass

At first sight the saving in V_C does not seem significant. Silica is inexhaustible, supplies of limestone are ample and soda ash can be produced from a naturally occurring mineral or by the chemical processing of salt or brine. These basic constituents of glass are not in short supply and their prices are low and stable. However, V_F, may be important in this context. Limestone quarries in the U.K. are situated in our National Parks. There is a significant amenity benefit to be credited to glass recycling if existing quarries have their lifetimes extended and new sites are not developed.

Process energy savings can be credited to glass recycling. There is a 2% energy saving for every 10% of cullet introduced into the furnace. Green glass with a 90% cullet content yields an approximate energy saving of 25%.

(c) Ferrous and non-ferrous metal

Neither iron ore nor bauxite supplies are limited on a global scale. Process energy savings are very high for recycled aluminium. If aluminium cans are being smelted in a reverberating furnace, the energy requirement is no more than 8.72 million Btu. This is under 4% of the 244 million Btu necessary to produce ingot from primary ore. At the national level, secondary aluminium smelting is less energy intensive than primary aluminium smelting. S_{St} is somewhat problematic, in terms of relying on householders to be able to consistently separate ferrous, mixed and aluminium cans in source separation schemes.

(d) Plastics

S_{St} is a major constraint on plastics recycling. Identification of the recyclable and non-recyclable plastics is not straightforward and source separation schemes would need to be targeted at a limited number of products (e.g. PET bottles) which are easily recognizable. Transportation also poses difficulties in cost terms. Mechanical separation plants have so far failed to separate both plastics and aluminium adequately.

S_D is perhaps more significant in the case of plastic than for paper and glass. The plastics component of municipal solid wastes is increasing very rapidly and it is non-biodegradable. This means that it causes particular problems (uneven settlement or voids) in landfill sites.

6.9 EVALUATING INDIVIDUAL RECYCLING SCHEMES

6.9.1 Source separation schemes

Fig. 6.2 shows the factors that have to be quantified and valued in any social cost benefit analysis of an individual source-separation scheme. An analysis of twenty or so schemes operating in OECD countries during the 1970s and early 1980s showed that a large proportion of them did represent economically viable options (Turner, 1983; Turner & Thomas, 1982).

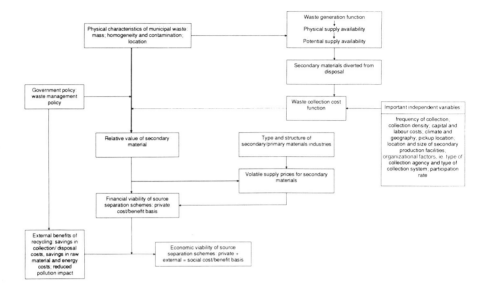

Figure 6.2 Social cost-benefit approach to source separation recycling schemes (adapted from R.K. Turner, 1981).

In the late 1980s in the U.K., there were source-separation schemes operating in at least 15 areas (Mansell, 1987). Three basic categories now exist:

1. drop-off point schemes, which rely on the public being willing to deliver reclaimable materials to accessible collection points, e.g. skips, bottle bins, igloos for paper collection. Good publicity, efficient administration and effective liaison with industry all appear to be important elements in the potential success of drop-off collections;
2. direct collection projects collections are made directly from waste producers, households, shops, offices and smaller factories. These projects are labour intensive and have in the past relied heavily on government job creation programmes;
3. direct collection with drop-off skips, direct collection is supplemented by bottle and can bank operations.

In 1988 the Community Programme Scheme was replaced by the Employment Training Scheme, marking a shift of emphasis away from the provision of work experience which benefits the community, to training through the acquisition of skill and confidence. A number of recycling difficulties have been inhibited by this policy change. Nevertheless, several new recycling ventures have been initiated recently, including the Recycling City Scheme. This scheme aims to establish model recycling facilities in four cities, the first scheme, in Sheffield, was launched in 1989.

6.9.2 Evaluating a resource recovery facility (mechanical separation)

For both the financial and wider economic evaluations, there are three distinct areas in which costs and benefits must be established if the evaluation is to be made correctly:

1. the actual recovery system and the related costs of handling, processing and transporting the waste and the recovered material or energy;
2. the alternatives for handling and disposing of waste and any savings that may result from recycling part of the arisings;
3. the nature of the material or energy recovered and its value by reference to virgin alternatives; and the net pollution impact associated with the alternative options.

Two basic accounting methods are displayed below in the context of a hypothetical resource recovery facility (Table 6.1).

Table 6.1 Resource recovery facility

Financial accounting	*Economic accounting*
1. Capital cost (fixed)	1. Capital cost
2. Interest of capital	2. Operational cost $(S_{Kt} + S_{St})$
3. Operational costs	3. Cost of residual waste disposal $(RS_D), + (RS_{DP})$
4. Cost of residual waste disposal	4. External costs, pollution etc. (S_P)
5. Total costs	5. Total social costs
6. Revenues	6. Revenues
7. Net revenues or costs	7. External benefits, $(V_F + S_D + V_P)$
8. Net profit or loss per tonne	8. Net social benefits or costs
Discounted cash flow over time	9. Net economic return or loss per tonne [Discounted net present value]
9. Compare with net cost of alternative disposal option i.e. landfill	10. Compare with net social cost of alternative disposal option i.e. landfill

The Social Cost-Benefit Analysis of the waste management option, incorporating a resource recovery facility, would need to be undertaken on the following basis (given data that are currently available in a monetary form) calculate the Present Value of Option's Net Benefit, NPV given by

$$NPV = \sum_{t=0}^{t=T} \quad K + \frac{B_t - V_{c_t}}{(1 + r)^t} \, ,$$

where, K = capital costs; B_t = benefits derived over time; V_{c_t} = operating costs over time; r = discount rate; and T = project lifetime; NPV = net present value of net disposal costs.

In more detail we have, on an annual basis:

$$NPV = \sum_{t=0}^{t=T} \quad K + \frac{P_r \cdot Q_r + (C_d \cdot Q - C_d) - VC \cdot Q - S_P}{(1 + r)^t} \, ,$$

where, Q = quantity of waste throughput; K = plant capital costs; $P_r \cdot Q_r$ = revenue from reclaimed materials; $C_d \cdot Q$ = total disposal costs (including collection and transport) C_c; C_d = residual waste disposal costs; $VC \cdot Q$ = plant operating costs; NPV = net present value of net disposal costs; and S_P = pollution costs.

It is likely that the 'demonstration' mechanical separation plants operating in the 1970s and early 1980s in the U.K. (e.g. Doncaster plant and Newcastle plant) would not have passed the economic cost benefit test. The £4.5 million Byker plant in Newcastle is still in service today (1990) but is producing well below its design capacity for waste-derived fuel output and resembles merely a waste transfer station. The more comprehensive Doncaster plant has been mothballed. The mid- to late-1980s have seen some renewed interest in energy from waste plants, and a number of ventures are currently under test at various locations in the U.K.

6.10 COMPREHENSIVE ENVIRONMENTAL APPRAISAL

Society has become increasingly aware of potential damage risks to the general quality of the ambient environment, and now demands that waste disposal and/or resource recovery options be both cost-effective and environmentally sound. Increasing concern has, for example, been expressed about the uncontrolled migration of explosive and other gases from landfill sites containing biologically active waste. The suspected escape into aquifers of chlorinated organic solvents originating from 'co-disposal' waste landfilling practice has also raised questions about relying unduly on the absorptive capacity of unsaturated zones underlying such 'dilute and disperse' sites.

In the future greater emphasis will have to be placed on engineered treatment before disposal of waste. The type and magnitude of emissions to the air via waste incinerators and refuse-derived fuel facilities also raise complex issues about pollution control costs and potential pollution risks.

In principle, wherever it is possible to quantify and value in monetary terms the impact of an option (landfill, RDF, composting or incineration) on the environment, an economic assessment will be sufficient to indicate environmental acceptability. However, a range of environmental impacts (biophysical and social) may not be amenable to monetary valuation. In these cases, environmental impact assessment techniques and/or multi-criteria evaluation techniques will be required, to quantify as far as is practicable the predicted effects and risks.

The concept of economically sustainable development has generated a great deal of debate, especially since the publications of the Brundtland Report written by the United Nations' World Commission on Environment and Development (Brundtland Commission, 1987). The economic meaning of sustainability has recently been examined (Turner, 1988; Pearce *et al.*, 1989; and Pearce & Turner, 1990). Essentially, sustainable economic development requires that the well-being of today's generation should not be increased at the expense of that of future generations. The next generation must have at its disposal at least the same productive potential as this generation.

A more formal interpretation of the sustainability requirement is that this generation should ensure that it passes on to the next generation a stock of assets no less than the stock it has inherited. The environment, deposits of natural resources, the ozone layer, the waste-assimilative capacities of rivers and oceans, etc. must be seen as a vital part of the stock of assets. During the process of economic development a degree of substitution is possible between man-made capital (machines and infrastructure, human skills and knowledge) and environmental capital. Some 'critical' environmental capital (ozone layer, store of biological diversity, etc.) is not, however, substitutable and must be conserved. It is vitally important that environmental assets are valued correctly before substitution processes take place. We need to price the environment as far as is practicable. Monetary valuation has limits and will require supplementation by environment impact assessment, carried out in quantitative but non-monetary terms.

A sustainable approach to waste management therefore will require a more integrated approach to the system and its evaluation. The term environmental audit has gained much prominence in recent years. A given municipal solid waste system option could initially be deemed efficient or socially worthwhile, in terms of its overall demand on natural systems. But this in itself is unlikely to be a sufficient evaluation procedure. Thus some reductions in the demand for natural resources and environmental services generated by a waste management option involving materials/energy recovery, can only be achieved by increasing the demand for non-environmental inputs, e.g. man-made capital. This increased capital demand may well then make its own demand on environmental inputs and also generate wastes.

The overall impact of, for example, refuse incineration is difficult to assess. Hydrogen chloride and dioxin emissions from incinerators have been the subject of much scientific debate. The quantities of potentially toxic emissions are low, but the pathways through the environment are complex and the toxicology of the chemicals involved is not well understood. These potential environmental risks, plus the capital and operating costs of the incinerator, have to be balanced against waste volume reduction savings and any energy recovery benefits.

Evaluation based on the economic criteria of social (private plus external) cost-effectiveness, and sustainability, involves three sequential procedures:

1. description or the technologies in question;
2. identification and measurement, as far as is practicable, of the overall economic/environmental impact of the technology and related activities. Input requirements, labour, capital, materials and energy, need to be quantified alongside environmental indicators. Amounts of pollutants emitted/discharged to the environment must be related to the quality (attributes) of the receiving environment;
3. weighting procedure, to establish the relative importance of the various disaggregated indicators. All inputs and pollution impacts must be priced in some way or another. The prices actually applied in the evaluation have to reflect accurately the social value of the resources concerned.

The pollution emissions or discharges index, for a given environmental receiving capacity, presents a complex evaluation task. Two basic procedures may be used. The first is to utilize whatever prices are available; the second relies on environmental impact (EIA) techniques and multi-criteria evaluation methods and techniques.

(a) Economic pricing

In this approach 'shadow' prices may be obtained via investigation of surrogate (proxy) markets for the impacts in question. Much progress has been made during the last decade or so in the field of economic valuation of environmental assets (Pearce & Turner, 1990). The bulk of the research work has sought to derive 'willingness to pay' monetary measures of environmental benefits. Nevertheless, the state of the art in this valuation field is far from being completely satisfactory. While it is possible, in principle, to categorize all the elements in the total economic value (use value + option value + existence value) of environmental resources, it is not possible to empirically measure them all.

Non-use values, option and existence values, in particular, are much more difficult to estimate. Furthermore, impact 'prices' should evaluate the impact (damage costs) of all pollution emissions or discharges, and this means that they will relate to the concentrations of pollutants, not the levels of emission or discharge. Environmental damage functions are often very complicated phenomena, especially when synergistic effects are present. The functions also vary with the different assimilative capacities of given environmental media at different locations. Random climatic variables sometimes further complicate the damage. On top of these physical uncertainties, there are social uncertainties in terms of the communication of information on, and public perception of environmental hazards. The environmental assessment procedure then is likely to have to encompass risk-benefit trade-off decisions based on 'expert judgement' values. This is especially the case where environmental changes are not amenable to specific measurement.

(b) Environmental impact assessment (EIA)

The EIA methods that were first developed (simple checklists, direct impact matrices and networks) concentrated almost exclusively on biophysical impacts expressed in non-monetary terms (Walthern, 1988). They were also mainly related to proposed developments on 'greenfield sites', and the likely significant impacts on the existing natural environment. More recently, methods have been developed which attempt to include a variety of socio-economic impacts, as well as biophysical impacts. It is also possible, in principle, to apply EIA techniques to operational plants which, by definition, will have already affected the local environment. In this context, impacts have to be judged against some 'acceptable' ambient environmental quality background. A minimum requirement for all waste management facilities is that their operations should be in compliance with the relevant statutory regulations. All EIA methods begin with the construction of an environmental inventory compiled from a checklist of environmental attributes representing the existing state of the local environment or exogenously determined 'acceptable' quality states.

Although checklists and simple matrix methods are useful gross screening devices, and also present an easily comprehended visual display, they are limited techniques. If alternatives are being appraised, there is always the temptation to score alternatives by adding unadjusted scores in the matrix. Such a procedure is incorrect and meaningless, given the ordinal nature of the scaled scores. Scoring systems even if technically correct should be supplemented by a decriptive assessment.

6.11 CONCLUSIONS

A fundamental switch in U.K. solid waste management philsophy from the established 'dispose, dilute and disperse' approach (via landfill), towards a 'recycle, concentrate and contain' approach now appears to be inevitable. It is the result of official EC pressure and mounting public environmental concern which have built up over the 1980s. There is now a governmental commitment to integrated pollution control encompassing BATNEEC and BPEO principles. These principles have been further buttressed by the growing level of support for the need to derive a sustainable economic development strategy.

Against this background the central question is, how can an appropriate municipal waste management system be chosen, given a wide range of available technical options? It has been argued that an economic appraisal of the basic recycle versus disposal conundrum indicates that current recycling activities should be extended. Waste management should be viewed as an integrated system encompassing, waste minimization, recycling and final disposal activities. Formally identifying the BPEOs within this system is, however, a formidably complex and resource-intensive task. Economic analysis can make an important contribution to the overall policy/options evaluation process, but it will also have to be supplemented with a number of non-monetary evaluation methods.

REFERENCES

Brundtland Commission (1987) *Our common future*. Oxford University Press.

Department of the Environment (1988) *Integrated pollution control: a consultation paper*. London: DoE.

Mansell, D. (1987) *A waste recovery project for Cambridge*. Cambridge: Friends of the Earth.

Pearce, D.W. and Turner, R.K. (1990) *Economics of natural resources and the environment*. Hemel Hempstead: Harvester Wheatsheaf.

Pearce, D.W., Markandya, A. and Barbier, E. (1989) *Blueprint for a green economy*. London: Earthscan.

Pessey, J. (1988) Market mechanisms of pollution control: 'polluter pays', economic and practical aspects. In *Sustainable environmental management principles and practice* (ed. R.K. Turner). London: Belhaven Press.

Royal Commission on Environmental Pollution (1988) Best practicable environmental option. Twelfth Report *Cmnd* **310**, London: H.M.S.O.

Turner, R.K. (1981) An economic evaluation of recycling schemes in Europe/North America. In *Progress in resource management and environmental planning* (ed. T. O'Riordan & R.K. Turner), vol. 4. Chichester: Wiley.

Turner, R.K. (1983) *Household waste: separate collection and recycling*. Paris: Organisation for Economic Cooperation and Development.

Turner, R.K. (ed.) (1988) *Sustainable environmental management: principles & practice*. London: Belhaven Press.

Turner, R.K. and Thomas, C. (1982) Source separation recycling schemes: a survey *Resourc. Pol.* **8** (1), 13–24.

Walthern, P. (ed.) (1988) *Environmental impact assessment: theory and practice*. London: Unwin Hyman.

Part Three

Dispersal

Chairman's introduction

P.F. Chester

The title of this session is misleadingly simple, with regard both to the content of the papers and to the nature of the disposal process to which it refers. The time-honoured way for mankind to dispose of its gaseous and liquid wastes has indeed been to disperse them to the winds or the tides, but this dilution and translocation is only the beginning of a chain of more complex processes which may include homogeneous and heterogeneous chemical reactions and the further adsorption, reaction or bioassimilation of the products at some intermediate or final destination.

Past ignorance of these complexities has given rise to a school of thought that would remove the dispersal option altogether from the portfolio of waste management techniques. Today's papers provide a glimpse of the alternative, adequate scientific understanding of the pathways to provide a basis for proper control of the disposal process.

Some of mankind's dispersed wastes, carbon dioxide and sulphur dioxide, for example, have their counterparts in nature and follow the natural degradation pathway. The goal here is to ensure that neither the pathway nor the receptor becomes overloaded. Enormous effort is being put into explicit physico-chemical modelling of acid rain and into even more complex modelling of the earth's carbon cycle. These models have yet to converge on concensus predictions. Meanwhile, the paper by Smith attempts to shed light on one aspect of the acid rain issue with a simple statistical approach.

Other wastes, transuranic elements and CFCs for example, have no natural analogue and their full environmental trajectory needs to be established. The paper by Woodhead & Pentreath is a good example of the use of radio-chemical tracers to determine the patterns of dispersion of soluble and insoluble materials in the marine environment and to reveal the differences in their takeup.

For waste disposal to ground, dispersal in the form of migration of the waste material or of a leachate may or may not be the desired outcome. In either case, the paper by Rae & Campbell show how far science has come in its ability to model such processes and to put the equally time-honoured practice of 'rubbish dumping' onto a sound basis. This progress assumes particular importance in the context of the North Sea Convention and the recent report of Sir Hugh Rossi's committee.

Waste reduction and recycling programmes, however effective, will not eliminate all waste. Absolute containment of all the residual waste would put a growing and unnecessary burden on society. Nature provides a substantial potential for physical, chemical and biological assimilation. The scientific challenge is to demonstrate the safe extent of this potential and to provide the means for policing its limits.

7

Possible future trends in acid rain in the U.K.

F.B. Smith

7.1 INTRODUCTION

Air pollution has existed on growing scales throughout the history of mankind. Although the industrial revolution marked a prolonged period of accelerating emissions, it is only in the past forty years or so that limited environmental effects of air pollution have become all too evident on scales of many hundreds of kilometres. Until the realization of these wider-range effects, it was almost inherent in man's inadequate thinking that the atmosphere and the earth were so vast that any amount of pollution could be injected into the air and that, apart from some local effects, this pollution would undergo enormous dilution, and the longer range consequences would be completely negligible. Legislation in the U.K., for example, was therefore aimed principally at minimizing the local effects, although some longer range benefits were also gained. The concept of the 'best practicable means (BPM)' was employed whereby industrial processes were obliged to employ the best economically viable technology to minimize the emission of noxious substances to the atmosphere. As part of this, the emitters were required to build sufficiently tall stacks to reduce ground-level concentrations and depositions to acceptable levels for both the local population and the environment.

Recognition of the wider-range effects has now put even greater emphasis on the need to abate emissions, especially from the larger emitters. These wider-range effects, frequently but not always, involve those associated with

The Treatment and Handling of Wastes
Edited by A.D. Bradshaw, Sir Richard Southwood and Sir Frederick Warner
Published in 1992 by Chapman & Hall, London, for The Royal Society
UK ISBN 0 412 39390 5, USA ISBN 0 442 31461 2

acid rain. Acid rain has both short-term and long-term effects on the environment. The effects seen today reflect to some degree the cumulative impact of depositions over many decades in the past. If emissions of those pollutants that directly or indirectly contribute to acid rain were to cease now, it seems probable that the environment would still be showing some adverse consequences well into the next century. Global emissions will not cease but are more likely to increase dramatically especially in the developing countries. Only in such developed areas as western Europe and North America is there a reasonable expectation that emissions will fall from their current high values.

In this paper the question is faced: how will levels of one of the dominant species in acid rain, namely sulphate, change in the U.K. between now and the first half of the next century? Levels of sulphur deposition will be influenced by:

1. changes in the emission of sulphur dioxide (and to a lesser extent by changes in the emission of other pollutants that become involved in the oxidation of sulphur dioxide to sulphate or its uptake into rain);
2. changes in those factors of the climate which lead to deposition; changes that in part at least arise from man's modification of the atmosphere through the build-up of greenhouse gases.

Both these factors are highly uncertain, and in consequence any conclusions reached within this paper must be considered speculative. Nevertheless, the question is an important one and demands consideration. It also permits some interesting new results from analyses of current deposition data to be set in a useful context.

The paper will consider the following points: (*a*) how are U.K. emissions of sulphur dioxide likely to change within the next 20 years; (*b*) how will the U.K. climate change as a result of the expected build up of greenhouse gases into the first half of the next century; and (*c*) how will these changes combine to affect the average sulphate depositions on the U.K.?

7.2 POSSIBLE CHANGES IN U.K. EMISSIONS OF SULPHUR DIOXIDE

Changes in the emission of sulphur dioxide depend on a large number of very uncertain factors: the future economic performance of the country, developments in energy-generation technology including flue-gas desulphurization, the availability and relative cost of different fuels with different sulphur contents, possible political decisions to limit emissions and major initiatives to conserve energy and to generate and use it more efficiently.

Predictions as to how these will work out in the next year, never mind the next 50 years, are bound to be very tentative. However, some pointers to the future are available. For example, the Government has made the decision to fit flue gas desulphurization equipment to three existing large coal-fired power stations and to all fossil-fuelled power stations to be built in the future.

The Watt Committee on Energy have fairly recently produced a report on 'Air pollution, acid rain and the environment' (Report No.18) in which predictions of emissions are made out to the year 2010, under certain assumptions. The Report gives a 'base-line' prediction, assuming no flue-gas desulphurization and a modest programme to build nuclear pressurised water reactors, which would give an annual emission of about 2.9 Mt of sulphur dioxide by 2010, down from a typical value of 3.4 Mt a^{-1} in 1985–87 when the Report was being prepared.

The situation has changed somewhat since the Watt Committee Report. The programme to build further PWRs has been suspended for the time being at least and Sizewell B may be the last to be completed for quite some time. This will result in a modest increase in emissions of about 0.3 Mt a^{-1}. The decision to fit FGD equipment on the otherhand should more than compensate for this by decreasing the emissions by about 0.4 Mt a^{-1} until a new generation of power stations have to be built to replace those coming to the end of their natural life. The programme to fit FGD equipment is proceeding only slowly and is reputed to be well behind schedule, while at the same time emissions have been creeping upwards. Recently, proposals for three new large coal-fired stations fitted with FGD to replace older stations have been cancelled. Nevertheless the Government, although not a signatory to the ECE so-called 30% club, has signed the EC Large Combustion Plant Directive thereby committing the country to the following cuts in sulphur dioxide and nitrogen oxide emissions, relative to 1980 values, (see Table 7.1). This Directive applies to existing power stations with a thermal output greater than 50MW.

In spite of all the recent adverse trends and cancellations, there will thus be considerable pressure on the Government and industry to meet the directive and to halve emissions of sulphur dioxide from current 1990 values within the first few years of the next decade. What happens beyond that is clearly uncertain, and so an emission value just less than 1 Mt S a^{-1} equal to half the current value will be assumed in this paper to apply into the time when the consequences of the greenhouse effect are fully manifested.

Table 7.1 Reductions in emissions relative to 1980 values entered into by signatories to the EC-directive

Emission reductions	*SO_2*	*NO_x*
by 1993	20%	15%
by 1998	40%	30%
by 2003	60%	—

7.3 POSSIBLE CHANGES IN THE U.K. CLIMATE

Greenhouse gases in the atmosphere are largely opaque to long-wave radiation and thereby alter the radiative balance of the planet. By far the most important greenhouse gas is water vapour. It, and some other naturally occuring gases, have made Earth habitable by raising the average surface temperature from about $-18°C$ to $+15°C$ as a consequence. Man's activities are expected to increase this mean surface temperature by perhaps some 1.5C or more in the next 40 years by increasing the airborne quantities of such greenhouse gases as carbon dioxide, ozone, methane, nitrous oxide and the chlorofluorocarbons (CFCs). The precise rise in temperature is still quite uncertain and is a matter of intense study at the present time. Several complex general circulation models have been developed which incorporate the main conservation equations for momentum, heat, moisture and mass. They differ in the way clouds are represented, in the complexity of the scheme by which the oceans and the vital role they play are included, and in a number of other parametrizations. These different models give rises in global temperature that vary over a factor of three. Also the forecasts these models give, using current conditions, diverge rapidly from reality in detail, but when extended over several years, approach the observed climatological state in a broad statistical sense, regardless of the initial conditions used.

The observed global temperature rise over the last century is consistent with the bottom end of the predictions of these models. Even so, puzzling differences are observed. The trend with time of the warming is different, as is the rise in polar-region temperatures relative to other areas which should be greater than actually observed so far. These throw doubt not only on the models as they are at present but also on whether the changes in climate to date are because of some completely different cause, and the consequences of the greenhouse gases are yet to materialize.

Focusing in on one small area, like the U.K., it is hardly surprising to find the models are in even greater disagreement. Only in one matter is there general accord: winter and spring rainfall is likely to increase (J. Mitchell, 1989, personal communication). When the atmospheric load of carbon dioxide has doubled from pre-industrialization levels, the models show an increase of about 0.75 mm per day average from November to May. The current national average is 2.48 mm per day, so the increase represents a rise of 30% during these months.

Although, clearly, this increase must be treated as uncertain, it will be used in this paper as an example of possible climate change for which the consequences can be explored, remembering that if at a later date a different climatic change becomes more probable, the method outlined herein could be used to derive the new deposition consequences.

7.4 THE CONSEQUENCE OF INCREASED RAINFALL TO DEPOSITION

An increase in rainfall may mean an average increase in rainfall intensity, an average increase in rainfall-duration during an event or an increase in the number of rain events themselves, or some combination of all three. If the models also agree that polar regions will warm up more than tropical regions, it might be supposed that the resulting lowered average latitudinal temperature gradient would not generally spawn more depressions that would eventually cross the U.K. than at present. The first two options would therefore seem more probable, but the conclusion must remain tentative. If an increase in rainfall intensity is dominant, then a decrease of the average concentration of sulphate in rain is to be expected, if other factors remain unaltered. This can be seen by the following somewhat simplistic argument. As stated earlier, intensity is related to raindrop size. The bigger the size the more the sulphate nucleus is 'diluted' by water. Consequently the concentration is less than for smaller drops. This implies a less than linear increase in sulphate deposition, the increase in rain being partly off-set by a decrease in the concentration.

If, on the other hand, the duration of rain increases, the deposition on the U.K. is likely to increase roughly in proportion since more of the U.K. emissions will be caught up in rain over the country.

Can we state which option is more likely? In principle, the models' results should be capable of giving the answer, but it is believed this study has yet to be made. The only other source of information is past data. The climate is constantly fluctuating and some periods in the past may be quite typical of average conditions in the future. This hypothesis is the basis of a series of small studies which will now be outlined,.

7.4.1 The relation between monthly rainfall anomaly and rain-days anomaly

The Meteorological Office has for many years published tables which give separately for England, Scotland, Wales and Northern Ireland the actual rainfall for each successive month (averaged over a large number of reporting stations) as a percentage of the long-term average for the appropriate month, and also the average difference in the number of days on which rain above a trace was recorded during the month taken from the norm. As expected, the data show a lot of scatter but when averaged a clear inter-relation emerges, as shown in Fig. 7.1. The higher the rainfall, the greater is the number of rain-days. Rain-days can increase either because more rain events occur or because the average duration of an event increases, increasing the probability the event will straddle part of two raindays. As simple arguments can show, a big increase in average duration is needed to

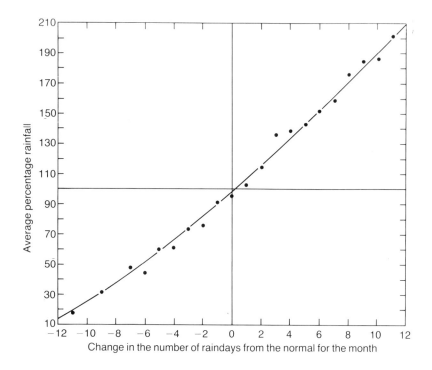

Figure 7.1 Monthly data giving the average anomaly in the number of rain days in the month at a number of observing stations in various regions of the U.K. are compared with the average percentage change in the rainfall. A six-year period has been analysed: 1983–88.

increase the number of raindays by just two days in a month. Also since the national average number of rain-days in a month is about 12, the data in Fig. 7.1 imply no significant change in rainfall intensity since a 25% change in rainfall, say, corresponds to a 25% increase in rain-days thereby requiring no change in intensity. More rain events seem the more likely cause for wetter than average months from this evidence.

7.4.2 Variation of sulphate concentration with rainfall

Inspection of any data set of concentrations of sulphate in rain versus rainfall amount reveals a marked tendency for the highest concentrations to be associated with low rainfalls. This has been studied by many workers. The connection is not altogether surprising for the reasons outlined earlier. However the relation may be due to another cause: in the U.K., for example, heavier rain is most often associated with westerly winds off the Atlantic carrying low sulphate concentrations, whereas easterly winds carrying more pollution tend to have only light rain or none at all. Source distribution is then correlated with rain intensity. To see whether this is the only cause, data may be analysed from a monitoring station in central Europe where upwind sources and resulting concentrations are much more uniformly distributed in direction. A clear example is given in Fig. 7.3 showing the results for a Czechoslovakian station CS1 where the concentrations have been averaged within back-trajectory sectors for given rainfall bands. The data are from two years, 1983 and 1985. A small sectorial dependence is evident, reflecting the variation of mean sulphate concentration in air with sector (see Fig. 7.2).

Jan. 1978–Oct. 1982

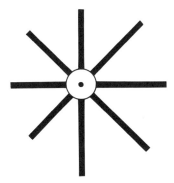

CS1 1 cm = 2μg m^{-3} S

Figure 7.2 The variation in the average concentration of sulphate in the air with the back-trajectory sector as defined in the EMEP programme at the Czechoslovakian site CS1. The variation is the least of all EMEP sites.

For U.K. stations a very similar variation of concentration with rainfall is found (see Fig. 7.4). The figure contains data from two U.K. stations which operate within the UNEP/WMO European Monitoring and Evaluation Programme (EMEP). The stations are at Eskdalemuir in southern Scotland (UK2) and at Stoke Ferry (UK4) in East Anglia. The data are for various timescales ranging from daily to 4-monthly. No division by trajectory sector is made. The daily data at UK2 were analysed for a subset (January–June 1985) of the 1984–87 period for which the monthly data were analysed and by chance have a somewhat higher mean concentration (0.75 mg l^{-1} compared with 0.54 mg l^{-1}).

The data are consistent with a simple relation

$$C/\bar{C} = 4(1 + Y)/(1 + 7Y),$$

where $Y = R/(MR_m)$; R, rainfall in the averaging period; R_m, average monthly rainfall; M, duration of the averaging period expressed in units of a month, and $\bar{C}$, the overall average concentration. Thus if a day has twice the average daily rainfall, $Y = 2$ and $C = \bar{C} \times 4 \times 3/15 = 0.8\bar{C}$. In reality there is a lot of scatter about this average and this must be partly due to dependence on the variable concentration of sulphate in the air at the time.

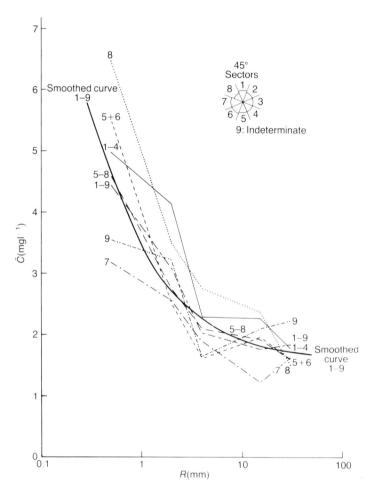

Figure 7.3 Daily data over two years, 1983 and 1985, have been analysed from CS1 to show the mean variation of concentration of sulphate in rain with rainfall amount. The thick continuous line is a smooth eye-fit to the overall variation. The other lines show the variation in various back-trajectory sector bands. Roughly 20% more sulphate per litre is observed in sectors 1–4 than in sectors 5–8 in agreement with the ratio of concentrations in air in the two directions as shown in Figure 7.2.

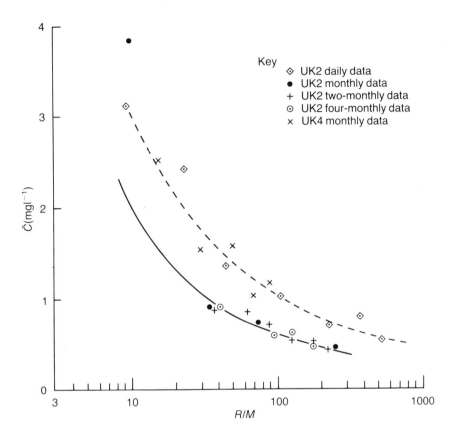

Figure 7.4 The variation of sulphate concentration in rain with rainfall at two British sites UK2 and UK4, for different sampling periods of duration M (in months). If allowance is made for the different average concentrations over the times from which the data were drawn, the data virtually fall on to a single curve, the empirical form of which is given in the text.

Fig. 7.5 shows the result of plotting C_R/C_A (where C_R is the concentration of sulphate in the rain and C_A is the concentration in the air) against rainfall amount, by using the Eskdalemuir daily data, on a log–log basis. In spite of the scatter, a marked trend is clearly discernible. If two rather extreme outliers are ignored (the points concerned are shown in parenthesis) the remaining points yield a regression line.

$$C_R = 0.51 C_A R^{-0.3} \text{ mg l}^{-1},$$

or deposition

$$D = C_R R = 0.51 C_A R^{0.7} \text{ mg m}^{-2}.$$

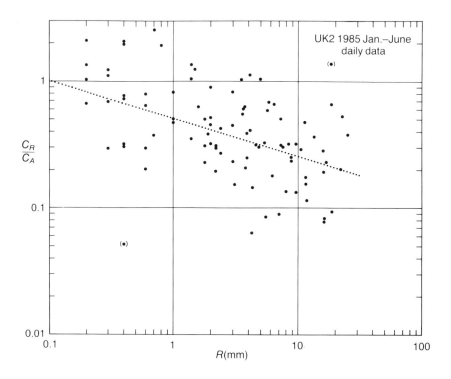

Figure 7.5 The ratio of sulphate concentration in rain to sulphate concentration in air at UK2 from January to June 1985 shows a marked trend with rainfall amount. If the two extreme outliers are excluded (they are marked by points in parentheses) the correction is $r = -0.5$ and a regression line is found whose form is given in the text.

Figs. 7.3–7.5 therefore point either to some average increase of rainfall intensity or to more prolonged spells of rain with increasing rainfall, in spite of the earlier indication that more rain events could supply the explanation. Perhaps we are forced to conclude that all these factors are present to some degree.

7.5 EXPECTED CHANGES IN SULPHATE CONCENTRATION DUE TO EMISSION REDUCTION ALONE

Extracting meaningful trends in pollution concentrations and depositions arising from changes in emissions is notoriously difficult and prone to abuse. Changes in year to year meteorological states can alone mask the true trends that are being sought as shown very clearly by Davies in the second Report of the U.K. Review Group on Acid Rain (1987).

Mylona (1989) has analysed changes in sulphur dioxide and sulphate concentrations in air detected at the EMEP stations in Europe during the period 1979–1986 and, while recognizing the many uncertainties involved, has noted an 18.9% reduction in observed concentrations of sulphur dioxide and a 15.5% reduction in the concentrations of particulate airborne sulphate, compared with an overall 16.4% decrease in European emissions over the 8 years.

To foresee the consequences of a reduction in U.K. emissions alone on average concentrations of sulphate in rain within this country, it is necessary to enquire how nonlinear the relation is likely to exist between them. Such an enquiry would involve a very careful study of micro-physical and chemical processes taking place in and around droplets within cloud, and turbulent mixing processes on a wide range of meteorological scales. Although this detailed study has not been completed, work in the U.S.A. within the National Acid Precipitation Assessment Program (NAPAP) is directed towards such a comprehensive model. In the meantime simpler simulations carried out in the U.K. indicate that nonlinear processes are in fact currently significant in major plumes from power stations and large industrial plants out to several hundred kilometres from source (Smith, 1987; Clark, 1987; Crane & Cocks, 1989).

It would be good to quantify the variation of concentration with emission levels by using past data. Unfortunately the year to year variations in mean concentration due to variability in meteorological conditions usually precludes this except in very broad qualitative terms. Most U.K. monitoring stations have only been collecting data since the late 1970s or early 1980s, which makes the determination of trends very uncertain. Only Eskdalemuir (UK2) has been operating continuously since the early 1970s. Its data show only a modest correlation with U.K. emissions ($r=0.70$) between 1973 and 1986. The scatter of the points is really too great to quantify a relationship with any sort of certainty. Nevertheless, an overall simple linear relation seems to fit as well as any other, implying that since the average concentrations at Eskdalemuir are already fairly low a halving of emissions may yield a virtual 50% reduction in concentrations in rain. However, more convincing results may sometimes be obtained by averaging results over several stations. Fourteen European stations in the EMEP programme have been in operation since 1974. An improved correlation of 0.864 is found between the average annual mean concentration of sulphate in rain and the annual emissions summed over the countries containing these stations. Further if it is assumed the mean concentration would go to zero if the emissions ceased (thereby neglecting small natural emissions), a simple quadratic expression can be fitted, which, however simplistic it might be, implies only a 30% reduction in concentrations if the emissions were halved.

A similar exercise applied to data from eight U.K. stations over a shorter period 1979–85 (the data come from Fig. 14 of the second Report of the U.K. Review Group on Acid Rain) gives a 36% reduction. Since five of these stations are relatively 'clean', it is not surprising the reduction is greater than for the 14 European stations: the lower the overall concentration the more linear the relation becomes.

This range in the percentage reductions to the overall pollution levels may be further affected by a possible, but hopefully small, response in the non-linear processes to the anticipated changes in overall climate. Whilst the US NAPAP Study may ultimately help to quantify such effects, they will not be addressed further in this paper.

7.6 IMPLICATIONS FOR FUTURE U.K. DEPOSITIONS OF SULPHATE

Earlier sections have indicated the following.

1. U.K. emissions of sulphur dioxide will fall relative to current values by at least half during the first decade of the next century if the Government honours its agreement to meet the EC-directive it has signed. What happens beyond that is extremely speculative.
2. These emissions will continue to dominate U.K. depositions since it is probable Western Europe will achieve similar targets, so the balance of home and overseas contributions will not radically change.
3. This reduction is likely to result in a 30–35% reduction in levels of sulphate in rain, if the climatology of meteorological conditions remain unaltered.
4. However, if the greenhouse effect is fully realized in broad accord with the consensus of climate model predictions (where such consensus exists) then by the time carbon dioxide concentrations have doubled, U.K. rainfall in winter and spring may have increased by some 30%. This doubling may be achieved in 40–60 years from now. No change in rainfall for the rest of the year will be assumed.
5. The increase in rain may be manifested in somewhat heavier rain, more prolonged rain and in more frequent rain events. Treating Eskdale-muir as typical in these respects, the data in Table 7.2 show the expected consequences. The first two changes in rainfall character will result in a drop in average concentration in rain, but because of the increase in total rain the resultant annual depositions will increase due to this cause alone by some 10%.
6. Combining (3) and (6) gives a 23–28% reduction in wet deposition of sulphate.

The conclusion must therefore be that, given all the highly speculative assumptions concerning future changes in U.K. emissions and changes in U.K. rainfall due to the greenhouse effect, the resultant decrease in sulphate wet deposition may be disappointingly small at around 25%, compared with the assumed 50% cut in emissions.

Table 7.2 Values of current average rainfall amounts at UK2 and estimated future values following a realization of the consequences of doubling carbon dioxide concentrations, together with the resulting changes in sulphate concentration in rain, for the months November to May. The depositions are found by taking the sum of the products of rainfall and concentration. The depositions from the months June to October are unchanged at 5.17 in these units. The ratio of future to current depositions is then 1.105, a 10.5% increase. C, 'current' average concentration of sulphate in rain for each month; $\bar{C}$, 'current' average concentration taken over the year; R, average rainfall for each month; $\bar{R}$, 'current' monthly rainfall averaged over the year; 'new' implies the state after the predicted changes in emission and climate 'old' refers to the 'current' state.

	N	D	J	F	M	A	M
$C/\bar{C}$ old	0.76	0.76	0.85	1.32	1.27	1.31	1.42
$R/\bar{R}$ old	1.56	1.40	1.40	0.68	0.74	0.48	0.72
$R/\bar{R}$ new	2.03	1.82	1.82	0.88	0.96	0.62	0.94
C_m (new)/C_m (old)	0.93	0.92	0.92	0.90	0.90	0.89	0.90

However, the total sulphur deposition has a significant contribution from dry deposition which is much more linearly related to the total national emission of sulphur dioxide. An estimate of the proportion of dry deposition to the total is shown in Fig. 7.6 which is reproduced by kind permission of Warren Spring Laboratory. It shows the proportion varies from around 10% in the remoter parts of Scotland to over 70% in some parts of eastern England, with an overall average of about 48%. This implies that once emissions are halved, the total sulphur deposition will be reduced by some 37%. However in some of the ecologically sensitive areas of England and Wales (e.g. the Pennines) where pollution levels are quite high and dry deposition currently accounts for only about 20–40% of the total, because of heavier rainfall in these areas, the reduction in sulphur deposition may only be 30–35%.

More recent estimates of dry deposition by D. Fowler (1990), based on field measurements, strongly suggest Fig. 7.6 actually overestimates the importance of dry to total deposition. These findings would therefore reduce the percentage reductions in sulphur deposition set out in the previous paragraph still further. A revised map taking on board Dr Fowler's results has appeared in the third Report of the U.K. Review Group on Acid Rain (1990).

Incorporating Fowler's conclusion into our findings may mean therefore that it may be difficult to detect from U.K. monitoring programmes clear trends in deposition resulting from changes in emission over the next two or three decades. Ultimately the principle of what is emitted into the atmosphere must be deposited somewhere, when combined with the fact that acidic depositions over significantly large areas currently exceed levels where ecological damage occurs, must lead to an increasing thrust to reduce continental-scale emissions, even though some areas may benefit more directly than others from these reductions, and greater benefit may be gained from reducing emissions in some localities than in others.

Even though climatic change cannot alter the overall global deposition it can alter its geographic distribution and, perhaps equally important, the sensitivity of the environment to the deposition. These interesting points will require further exploration.

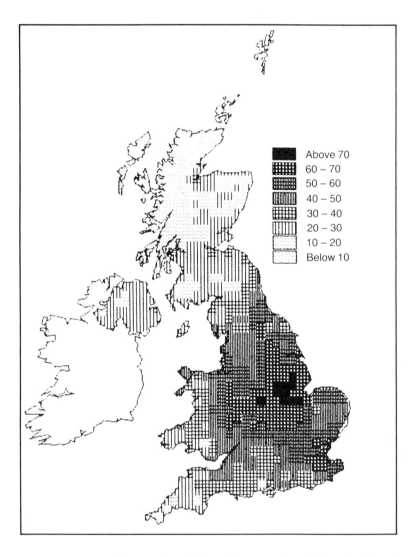

Figure 7.6 The estimated proportion of the total average deposition of sulphur from dry deposition. (Reproduced by permission of Warren Spring Laboratory.)

REFERENCES

Clark, P.A. (1987) The influence of the nonlinear nature of wet scavenging on the proportionality of long term average sulphur deposition. In *Interregional air pollutant transport: the linearity question* (ed. J. Alcomo, H. ApSimon and P. Builtjes), IIASA RR–87–20. Austria: Laxenburg.

Crane, A.J. and Cocks, A.T. (1989) The transport, transformation and deposition of airborne emissions from power stations. In *Acid deposition: sources, effects and controls* (ed. J.W.S. Longhurst). British Library Technical Communications.

Fowler, D. (1990) A submission to the U.K. Review Group on Acid Rain.

Mylona, S.A. (1989) Detection of sulphur emission reductions in Europe during the period 1979–1986. MSC-W Report 1/89.

Smith, F.B. (1987) The response of long-term depositions to non-linear processes inherent in the wet removal of airborne acidifying pollutants. In *Interregional air pollutant transport: the linearity question* (ed. J. Alcomo, H. ApSimon & P. Builtjes), IIASA RR–87–20. Austria: Laxenburg.

U.K. Review Group on Acid Rain (1987) Acid deposition in the United Kingdom, 1981–1985. Second Report. Stevenage: Warren Spring Laboratory.

U.K. Review Group on Acid Rain (1990) Acid deposition in the United Kingdom 1986–1988. Third Report. Stevenage: Warren Spring Laboratory.

Watt Committee on Energy Ltd (1988) *Air pollution, acid rain and the environment* (ed. K. Mellanby). Report No.18. London: Elsevier.

8

The migration of pollutants in groundwater

J. Rae and D.J.V. Campbell

8.1 INTRODUCTION

The United Kingdom has relied, almost exclusively, on landfill disposal as the ultimate route for a wide variety of wastes. The potential threat posed by leachates generated within such wastes has been recognized for a long time and, more recently, attention has also been given to the production and migration of landfill-derived gases. As a consequence there is considerable interest in research and development projects within the waste disposal industry and in the Department of the Environment, which supports many of these projects.

As our understanding of waste degradation processes; physical, chemical and microbial has increased, so has our knowledge of the interaction and variations in leachate and gas quality and production. This has allowed the development of improved legislation, monitoring techniques and operational practices to be introduced to meet higher standards of environmental control. In turn, the continuing practice of landfill disposal of most municipal, commercial and industrial waste can be viewed with better confidence.

8.2 RESEARCH PROGRAMMES

Legislation specific to waste disposal was first enacted in the U.K. in 1972 and quickly superseded by the Control of Pollution Act 1974. The DoE commitment to research into the topic began shortly after the 1972 Act with

The Treatment and Handling of Wastes
Edited by A.D. Bradshaw, Sir Richard Southwood and Sir Frederick Warner
Published in 1992 by Chapman & Hall, London, for The Royal Society
UK ISBN 0 412 39390 5, USA ISBN 0 442 31461 2

a multidisciplinary study on the behaviour of hazardous wastes in landfill sites. In addition to detailed leachate water quality monitoring in and around 19 pre-existing sites, experimental studies were set up both in the field and in the laboratory. They examined the fate of hazardous wastes co-deposited with municipal wastes, and the fate of leachates generated from such wastes as they migrated through different unsaturated strata into underlying aquifers. The final report (DoE, 1978) concluded that both the studies and previous experiences had shown that 'very few documented cases of significant groundwater contamination due to landfills have occurred' and that 'the research findings generally have confirmed this view and have now provided the basis for a better understanding of the natural processes involved'.

In the past ten years research has diversified considerably and about £15 million has been spent specifically on research into controlled waste management. Table 8.1 (Bentley and Gronow, 1989) gives a breakdown of the areas of research for financial year 1989–90 when DoE will spend about £1.3 million on landfill research, including £0.5 million on problems associated with landfill gas.

Table 8.1 Landfill research programme

	Number of projects	*Cost (£)*
Modelling, risk assessment and monitoring	7	375 000
Microbiology of leachate and gas	7	275 000
Mechanisms of migration and attenuation	11	325 000
Leachate management	7	300 000
Re-instatement of landfills	2	25 000

An interesting report on two of these programmes is given in Robinson (1989), a paper that presents detailed monitoring results from two major landfill sites in southern England. One, at Compton Bassett in Wiltshire, is a containment site where filling takes place in separate cells constructed of Kimmeridge Clay, from where leachate is extracted as it arises to be treated at an on-site plant and discharged. The other, Stangate East in Kent, is designed to operate on the 'dilute and attenuate' principle with leachate allowed to pass in a controlled manner through an engineered unsaturated zone beneath the base of the landfill where it is attenuated and purified before entering the saturated zone beneath. Data from both sites cover several years and give a detailed account of the development of methanogenesis with respect to both gas production and leachate quality.

8.3 LEACHATES

The implication of laboratory and field studies is that properly managed landfills are natural bioreactors where, in particular, microbial processes have a critical part to play not only in degrading organic matter, but in influencing conditions for retention of metallic species (e.g. by pH control) and the quality of leachates and gases generated. Similarly, such processes can occur within strata surrounding landfill sites to attentuate potential pollutants in leachates.

The biochemistry of landfills can be summarized and encapsulated in Fig. 8.1 (after DoE, 1986). Following waste deposition, compaction and placement of daily and, in due time, final cover materials, waste environments rapidly become anaerobic as oxygen is utilized by microbes with production of carbon dioxide. Anaerobic bacteria then become dominant and hydrolyse and ferment organic matter, degrading complex organic molecules to produce acids, ammonia and further carbon dioxide. During this stage in the

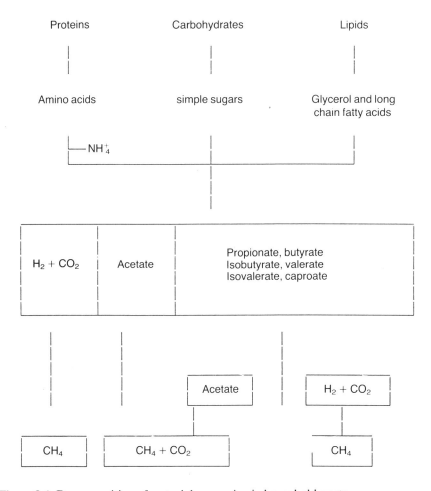

Figure 8.1 Decomposition of materials occurring in household waste.

process acidic conditions pertain with the pH falling to 5.5–6.5. Under such conditions many metal ions will be soluble. If this process is maintained, as can occur in poorly managed sites where rainfall infiltration is particularly excessive, the consequences will be continued production and eventual release of large volumes of highly polluted leachates (with high Biological Oxygen Demand (BOD), metals and ammonia). However, many wastes can themselves act as a buffer medium to control pH and can absorb liquids. If rainfall infiltration and leachate volumes are controlled, further degradation of organic acids occurs, ultimately by methanogens (strict anaerobes) to convert acetate to methane and carbon dioxide. Hydrogen gas, also generated during initial refuse fermentation, is also utilized by methanogens to convert carbon dioxide to methane and water. The pH will be maintained at or near neutral during this final phase of refuse degradation, resulting in *in situ* precipitation of, for example, heavy metals. These processes will also naturally occur in suitable strata beneath landfills to further 'clean up' leachates. Cation exchange with clay minerals in soils and rock may also assist in removal of any heavy metals remaining in leachates leaving the site. Other organics, for example, phenols, can similarly be degraded microbially within wastes and many inorganic metal sludges will be retained in the waste mass. There is of course a need to maintain good control over these processes, by sound operational practices. Apart from preventing excessive leachate production there may be a need to collect and treat externally, or by recycling, leachate volumes produced above and beyond those which can be successfully attenuated or diluted beneath the site. Increasingly, where geological or hydrogeological conditions are shown to be unsuitable, use of a site may only be permitted if complete containment is provided either by natural or artificial means. All containment sites must have provisions for external leachate treatment to preclude excessive build-up within sites (as should attenuation sites).

While discharge to sewers for external treatment has commonly been practical, the DoE have undertaken research on both aerobic and anaerobic methods for treating leachates. As has been indicated above, leachate quality will vary with time, progressively becoming less organically rich, with a BOD reducing from as high as 30 000–50 000 mg l^{-1} in the very early stage to less than 500 mg l^{-1} and with consequent reduction in metal contaminants as well. Ammonia tends to be the major pollutant of leachates from 'mature waste' undergoing stable degradation (but there are of course all phases of waste degradation occurring simultaneously under these conditions). Given the variability of leachate quality, and the increased expense involved in anaerobic treatment, aerated lagoons offer the best 'solution' before discharge of liquids to surface water courses. In the unsaturated zone beneath sites oxidation of ammonia can also occur beneath any anaerobic environment that becomes established. Iron, reduced and soluble under anaerobic conditions, will be precipitated out of solution during aeration, and will form complexes with other metals too. High strength leachates (high BOD) of low volumes may be recirculated back into wastes for further degradation. DoE have funded research to examine spray irrigation of leachates both to reduce volume by evaporation and to provide attenuation or degradation of contaminants in it. There have been shown to be some critical limitations to this technology both in terms of its effectiveness, and on human and animal health grounds.

Where natural low permeability materials (generally clays) are not readily available to provide partial or complete containment or where little or no unsaturated zone exists beneath sites, other options exist. An example is the engineered site in Kent mentioned earlier. Detailed instrumentation has been installed within its artificial undersaturated zone to monitor gaseous and liquid products to evaluate its effectiveness at attenuating leachates migrating through it.

Substantial databases of leachates derived from many landfill site sources, and types, have been accumulated during the course of various research studies and the DoE are currently evaluating such data within studies to review leachate quality; methods of leachate treatment; and impacts of co-disposal of hazardous wastes on leachates.

8.4 LANDFILL GAS

The ultimate product of organic matter degradation in sites is landfill gas, a mixture of gaseous components. Principal gases produced during steady-state conditions are methane (60–65%) and carbon dioxide (40–35%). There are many other trace gases too (more than 120 have been identified), as products of microbial, chemical or simple waste evolution processes, for example, from aerosol cans.

As both landfilling practices improve, and as sites have become fewer, but larger to allow regional economies of scale, so anaerobic gas production has increased, in comparison with former practices where aerobic degradation was dominant, and even waste burning was commonplace. During the early 1980s increased concerns about the risks and hazards of landfill gas were rising. There were incidents involving explosions in services adjacent to sites, excessive gas in or near to property and evidence of damage to vegetation. The Loscoe incident (Public Inquiry, 1988), and explosion that demolished a bungalow in Derbyshire, sharply focused attention on the problem. Landfill gas generated in significant quantities will create a positive pressure with respect to atmosphere and, with increased use of low permeability caps above sites, will tend to migrate laterally beyond site boundaries into permeable strata if and where no barrier has been provided and/or where gas is not released to the atmosphere. The DoE have given con- siderable attention to this problem, looking at methods to prevent gas migration and ensure its safe release, generally after flaring, to atmosphere, and formulating guidance on the control of landfill gas (HMIP, 1989). Studies are being undertaken to look at gas permeabilities of site lining and sealant materials, as well as techniques for monitoring and for gas abstraction either passively or actively pumped.

Some types of strata may well have gases (often methane) in them derived from other sources than from landfill. In evaluating the monitoring data associated with potential for landfill gas migration, it is sometimes necessary to distinguish gas sources. Various fingerprinting techniques have been developed, including use of gas chromatography (GC), C_{14} dating, C_{13}/C_{14} isotope ratios, and GC/mass spectrometry. This latter technique has particular applications for examining minor components present in landfill gas and assessing the extent to which components may be above acceptable TLVs or responsible for odour nuisance, a common complaint at or near landfill sites where gas emissions are not under control.

There is much interest in the commercial exploitation of landfill gas as an energy resource and, through the Department of Energy, many research and demonstration projects have been undertaken to further this technology. Projects range from gaining a better understanding of refuse decomposition processes and methods for enhancing gas production to improving our understanding of methods of gas recovery via gas wells or horizontal trenches. The largest field experiment, at Brogborough in Bedfordshire, includes six 15 000 tonne 20m-deep test cells in a commercial landfill site. Several demonstration projects have been instigated to look both at the efficiency and problems of various recovery plant designs and end uses. Today gas is being used in boilers, cement and brick kilns, for electricity generation via gas or diesel engines, and turbines, as well as in processes to 'purify' landfill gas to produce 'pure' methane.

8.5 ATTENUATION AND MODELLING

When waste is landfilled there are a number of processes that can reduce its polluting potential, including dilution and dispersion as well as the chemical and microbial processes discussed above. When leachate migrates with groundwater from a site into surrounding soil and rock still more processes come into play. These may be of a physical, chemical or biological nature and to the extent that they reduce pollution are usually called attenuation. The prevailing hydrogeology, relation of saturated and unsaturated zones and the geochemistry of the surrounding strata all play a part. Their combined effect can often result in a reduction in the concentration of pollutants by several orders of magnitude, but it must be admitted that many of the mechanisms of attenuation process are not well understood. The processes usually considered are listed in Table 8.2 (after DoE, 1986).

These processes all interact with one another in a very complex manner. For example, the pH and redox potential at a particular place in the

Table 8.2 Attenuation processes

Physical	Absorption, adsorption, filtration, dilution, dispersion, diffusion
Chemical	Acid-base interactions, oxidation, reduction precipitation, ion-exchange, complex ion formation colloids
Biological	Aerobic and anaerobic microbial degradation

groundwater flow will depend on both the leachate and geochemistry of strata and will in turn govern much of the chemical and microbial activity at that place.

One common method of testing our predictive understanding of such processes, and especially of their combined effects, is to use mathematical models. This is a large subject in its own right and only a brief outline can be given here.

Perhaps the simplest models used in this context are of the 'tank' type whereby the pollution species are attributed to conceptual 'tanks' with different characteristics and are allowed to exchange between 'tanks' at assigned rates. Sometimes the attribute is chosen to model the actual geology, the 'tanks' represent strata and the exchanges the water-flow balance. Sometimes the attribute is more conceptual representing perhaps different geochemical environments. The great merit of such models is they can be made very simple and dependent on just a few chosen parameters. The drawback is that these parameters are usually very indirectly related to our understanding of the physics and chemistry of the situation.

The next level of model, widely used in the 1970s, describes the change in concentration of some pollutant species as it moves along a one-dimensional flow path with a prescribed velocity. The equations that govern this change in space and time are partial differential equations in time and one space dimension as independent variables, and have the general structure:

$$R \frac{\delta C}{\delta t} + V \frac{\delta C}{\delta x} - \frac{\delta D}{\delta x} \frac{\delta C}{\delta x} = S \tag{8.1}$$

Here C is the concentration, x and t the space and time variables, and V the prescribed flow velocity. All the chemical and physical sorption effects are rolled up in a retardation factor R, all diffusion and dispersion effects in D and all sources and sinks, including chemical changes, in the term S. Although more explicitly reflecting the physics and chemistry of the situation, it is clear that models of this type are still some steps removed from a detailed understanding. Models of this type, when it comes to calculating answers, are usually converted into numerical models by one of a number of well-known techniques. It is worth commenting in passing that this step is not always straightforward. Because retardation factors R and source terms S can differ by orders of magnitude for different species, a mix of pollutants will contain many timescales and 'stiffly stable' numerical methods need to be used (Byrne and Hindmarsh, 1975). Also, if the ratio of the dispersion term D to the advection term V gets too large for the numerical grid spacing, numerical instability can result in many models. This has implications for the cost, complexity and accuracy of the model and can be particularly troublesome in the case of gas flows.

In the course of the 1980s rather more elaborate models have become common. This in part due to advances in computing power, in part, because of increasingly difficult circumstances for oil and gas production, but in no small way because of the need to assess in detail the safety and environmental consequences of the burial of radioactive waste. There are many common features to the land burial of active and non-active wastes and each side of the industry has much to learn from the other. It is typical of these recent models to take the calculation in several stages: first to calculate the groundwater flow: next to look at the motion of the pollutant, in solution or otherwise carried along, taking account of diffusion, dispersion, local chemistry, sorption and so on, at a not-too-detailed level of parametrization; then to perform side calculations, sometimes of considerable sophistication, to model these processes in detail: finally to iterate between the preceding two stages.

The groundwater flow calculation is usually done in terms of Darcy flow in a porous medium with the equations being solved numerically in 1, 2 or 3 dimensions as appropriate over the region of interest (which with modern computing power can be quite extensive). The output from this is a set of flowpaths and rates of pollutant transport. In some rock-strata flow may be mainly in faults or discrete fractures and then the question arises as to the relation, if any, with porous medium flow. This is an area in which steady progress has been made in recent years.

With the aid of a CRAY-2 computer the Harwell fracture network model NAPSAC has modelled a flow in up to 10 000 fractures in three dimensions, discretized by up to 10 million elements. This type of model has benefited greatly from the extensive experimental work being carried out on the fractured granite system at the Stripa Mine in Sweden.

The pollutant transport equations are usually multidimensional analogues of equation (8.1) with variable values for the coefficients and sometimes other processes included such as rock matrix diffusion. They continue to have the same potential numerical problems mentioned above in connection with equation (8.1). The side calculations of the transport coefficients are sometimes very detailed. For many systems, for example, the retardation to migration provided by sorption is very significant and the use, as is common, of a simple linear equilibrium sorption model needs to be justified. There are now more complex representations of sorption, which can be accommodated in 'chemistry and transport' coupled models. The CHEQMATE code developed at Harwell is an example of this.

It models the evolution of spatially inhomogeneous aqueous chemical systems characterized by simultaneous chemical reaction and ion migration processes. CHEQMATE iteratively couples these two processes, so that local chemical equilibrium is maintained as the transport processes evolve. Ion migration is included by diffusion, dispersion, electromigration (to maintain local charge balance) and advection. The chemical part of the code calculates the equilibrium water chemistry for a particular chemical inventory, drawing on a database of thermodynamic data. It includes a mineral-accounting procedure, so that solid phases may be added or removed from the system as precipitation or dissolution occurs. Nonlinear sorption processes can be modelled within the chemical equilibria part of CHEQMATE and the reactions between the aqueous species and mineral surface sites take into account electrostatic interactions.

These state-of-the-art models are now being applied more widely but are not without their problems. There are still difficulties with fracture flow, gas flow and the undersaturated zone, for example. Above all, perhaps, is the difficulty in acquiring sufficient adequate data both to validate and run the models for realistic situations.

8.6 CONCLUSIONS

It is clear that despite its long history, the practices and processes of landfill disposal are complex to follow in detail and that new legislation and public conern will require us to understand them better. It is perhaps particularly important to address the interactions of the various aspects of landfill disposal. The DoE issued in 1986 an article entitled 'Landfilling wastes', which provided advice on good landfilling practice. This paper distilled the experience of some 90 papers on aspects of good practice and the expertise of many practioners and researchers. It emphasized that while there were many disciplines necessary to provide a sound basis for landfill development, there was a need for them to interact with each other. In particular, there is a need for greater understanding of the interactions between leachate production and generation and management of gas.

It is probable that landfill sites will increasingly be required to conform to a 'containment' philosophy in the future, particularly to ensure that landfill gases do not migrate beyond site boundaries. The 'disperse and attenuate' philosophy as commonly adopted in the past is still considered technically sound in the right geological setting and will still have a part to play in protecting the environment. There is no validity in the notion of containment for all time. While containment may be preferred in the most active phases of waste degradation, the long-term controlled reassimilation of waste contaminants back into the environment must occur eventually.

In the future, landfill processes, and hence leachate and gas quality, could be markedly different, as and when the nature of the deposited wastes changes. Decreasing landfill space, conservation of raw materials, recycling and reuse of waste products will all provoke such changes. In the meantime continual monitoring of liquid and gaseous effluents is needed to improve our understanding.

ACKNOWLEDGEMENTS

We thank our colleagues at AEA Technology for providing background material for this paper; in particular, we thank Chris Dent and Phil Tasker.

REFERENCES

Bentley, J. and Gronow J.R. (1989) Environmentally acceptable landfilling in the UK. In *Proceedings of the 2nd International Landfill Symposium*, A III, pp. 1–13. Sardinia.

Byrne, G.D. and Hindmarsh, A.C. (1975) *ACM Trans. Math. Softw.* **1**, 71–96.

Department of the Environment (1978) *Cooperative programme of research on the behaviour of hazardous wastes in landfill sites.*

Department of the Environment (1986) *Landfilling wastes.* Waste management paper no. 26.

Her Majesty's Inspectorate of Pollution (1989) *The control of landfill gas.* Waste management paper no. 27. London: H.M.S.O.

Public Inquiry (1988) *Report of the non-statutory public inquiry into the gas explosion at Loscoe, Derbyshire, 24th March 1986.*

Robinson, H.D. (1989) Development of methanogenic conditions within landfills. In *Proceedings of the 2nd International Landfill Symposium*, XXIX, pp. 1–9. Sardinia.

9

The dispersal of radionuclides in the sea

D.S. Woodhead and R.J. Pentreath

9.1 INTRODUCTION

Artificial radionuclides represent just one of the many classes of waste material discharged, or dumped, into coastal waters. Because their presence in this environment has the potential for increasing the radiation exposure of the general public, either through the consumption of contaminated seafoods or through external exposure from contaminated sediment on beaches, such discharges have been carefully controlled (see, for example, Hunt (1985a)).

Notwithstanding the generally low level of the authorized discharges, the consequent presence of the artificial radionuclides in the sea has been of particular scientific interest because it has provided a potentially valuable means of tracing and quantifying the many, and inter-related, physical, chemical and biological processes taking place in the marine environment. For a number of reasons, therefore, the behaviour and distribution of the artificial radionuclides in the sea have, perhaps, been subject to as much detailed study as those of any other single class of contaminant.

The utility of the artificial radionuclides derives from a number of considerations. The majority of controlled inputs to the marine environment have a reasonably well-defined source term. The phenomenon of radioactive decay provides two useful properties: a given radionuclide can usually be unambiguously detected at very low gravimetric concentrations; and under appropriate conditions it can provide a timescale which may yield information concerning the rates at which various processes take place.

The Treatment and Handling of Wastes
Edited by A.D. Bradshaw, Sir Richard Southwood and Sir Frederick Warner
Published in 1992 by Chapman & Hall, London, for The Royal Society
UK ISBN 0 412 39390 5, USA ISBN 0 442 31461 2

Lastly, the different radio-elements present in the discharges can exhibit a wide range of geochemical properties.

9.2 THE SELLAFIELD DISCHARGES

Since the processing of irradiated uranium commenced at Sellafield (then Windscale) in 1952, a variety of radionuclides, with a wide range of half-lives and markedly dissimilar geochemical properties, has been discharged, under authorization, into the north-east Irish Sea. In the early years the total quantities of radionuclides discharged were small, but with the inception of the programme of commercial production of electricity from nuclear power (in 1962), and the associated reprocessing of the uranium fuel, the releases increased.

There are two elements present in the effluent, caesium and plutonium, which show contrasting behaviour in the marine environment due to their differing chemical properties. Caesium, represented by the nuclides Cs-134 and Cs-137, is essentially conservative in sea water, i.e. the element shows relatively little affinity for particles as indicated by the distribution coefficient defined as:

$$K_D = \frac{\text{concentration of nuclide in particulate (Bq kg}^{-1})}{\text{concentration of nuclide in filtrate water (Bq l}^{-1})}.$$

Thus, with a relatively low K_D of 3×10^3 (IAEA, 1985), only a small proportion of the input is incorporated into settled sediments, while the greater part remains in solution in the water column where it provides a tracer of water movement. The second element, plutonium, is represented by isotopes with atomic mass numbers of 238, 239, 240 and 241. The decay characteristics (the α-particle emission energies) of Pu-239 and Pu-240 are not sufficiently different that they may be quantified separately by radiometric analysis; Pu-238 is also an α-emitter but may be assayed separately. Pu-241 is a β-emitter which is more difficult to assay and will not be considered further here. Plutonium has a high K_D value of 10^5 (IAEA, 1985) and the greater part of the input is rapidly scavenged to the sea bed by the resuspension-settling cycle of fine particulates. The characteristics of the radionuclides are summarized in Table 9.1.

The variations of the discharges of the nuclides with time are shown in Figs. 9.1 and 9.2. The values for Cs-134 and Cs-137 are derived from data given in the annual reports of British Nuclear Fuels (BNFL, 1984–89) and a summary report prepared by the National Radiological Protection Board (Stather *et al.*, 1986). The caesium discharges began building up in the late 1960s with the greater fuel throughput generated by the expanding nuclear power programme. There was a temporary halt to fuel reprocessing in 1974 which resulted in an increase in the fuel storage time in the cooling ponds, leading to greater corrosion of the magnox fuel cladding and a consequent

Table 9.1 Radionuclide characteristics and Sellafield discharges to December 1988

Radionuclide	Half-life (years)	Major radiation emissions			Practical limit of detection		Sellafield discharges	
		Type	Energy[a] (MeV)	Yield	(Bq)	(g)	Integrated annual input (TBq)	Decay-corrected environmental inventory (TBq)
Cs-134	2.06	β	0.024 0.210	0.27 0.70	5×10^{-2}	10^{-15}	5825	143
		γ	0.605 0.796	0.98 0.85				
Cs-137 plus (Ba-137m daughter)	3.00×10^{1}	β	0.173 0.425	0.95 0.05	5×10^{-2}	2×10^{-14}	40910	30980
		γ	0.662	0.85				
Pu-238	8.77×10^{1}	α	5.46 5.50	0.28 0.72	10^{-5}	2×10^{-17}	143	129
Pu-239	2.41×10^{4}	α	5.11 5.14 5.16	0.11 0.15 0.74	10^{-5}	4×10^{-15}	588	588
Pu-240	6.54×10^{3}	α	5.12 5.17	0.27 0.73	10^{-5}	10^{-15}		

[a] For β-particles the maximum emission energy is given.

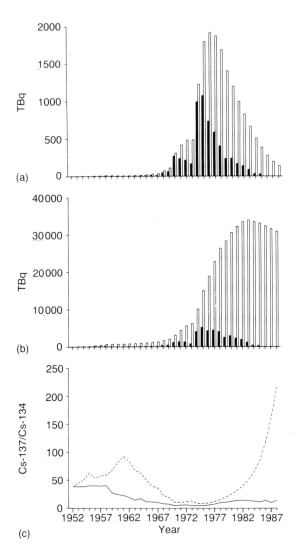

Figure 9.1 Discharges of Cs-134 (*a*) and Cs-137 (*b*) from Sellafield to the north-east Irish Sea; the filled bars denote the annual discharge and the open bars denote the decay-corrected, cumulative environmental inventory. (*c*) The Cs-137:Cs-134 ratio in the effluent (full line) and for the decay-corrected cumulative environmental inventories (dashed line).

increase in the release of caesium nuclides to the pond water. These circumstances produced the sharp rise in discharges in 1974 and a broad peak extending to 1978. Chemical treatment of the pond water to inhibit corrosion and the use of zeolite skips to reduce the caesium concentrations produced a gradual decline in discharges until 1985 when the commissioning of new effluent treatment plant (SIXEP) resulted in a further major reduction. The Cs-137:Cs-134 activity ratio is shown in Fig. 9.1c for both the annual discharges and the decay-corrected environmental inventory. This ratio is a potentially useful means for 'ageing' the caesium signal in environmental materials although such use is dependent on the assumptions of constancy in the source term and the dominance of advective transport over diffusion in the environment (Jefferies *et al.*, 1982; Hunt, 1985b).

The data for Pu-238 and Pu-239+240 discharges to sea (Fig. 9.2) for the period 1978–88 were derived from the same sources (BNFL, 1984–89; Stather *et al.*, 1986). Before 1978, the input was given in terms of total Pu-α, i.e. Pu-238 plus Pu-239+240, although measurements on environmental materials indicated that varying proportions of the nuclides were present in the effluent (Hetherington, 1978). Estimates have been made of the relative proportions of these nuclides in the earlier discharges by using measurements made on samples taken from long sediment cores collected in 1988 from a disused dock in Maryport Harbour (Cumbria). The dock was dredged in the early 1950s but was not used commercially. It rapidly silted up and in the process appears to have incorporated an undisturbed record of the Sellafield discharges (Kershaw *et al.*, 1990). These measurements yield estimates for the discharges of the individual nuclides for the time period 1959–77; for the period from the commencement of discharges in 1952 to 1958, it has been assumed that the Pu-239+240:Pu-238 activity ratio had a value of 25. The inputs of Pu-239+240 to the north-east Irish Sea increased steadily through the 1960s to reach a peak in 1973; since then, there has been a more or less consistent decline. The Pu-239+240:Pu-238 activity ratio is given in Fig. 9.2c for the annual discharge and the cumulative environmental inventory and these data provide a potential means of estimating the age of the plutonium content of environmental samples (again, subject to the caveats noted above for the Cs-137:Cs-134 ratio). The Cs-137:Pu-239+240 activity ratio (Fig. 9.8a) also provides a useful context within which to interpret the corresponding ratio in environmental materials.

9.3 ENVIRONMENTAL DISTRIBUTIONS

9.3.1 Caesium-137 in sea water

The distributions of Cs-137 in the waters of the NW European continental shelf are monitored regularly for the purpose of making a radiological assessment of the wider impact of the Sellafield discharges. Figs. 9.3a–c show the distribution in U.K. coastal waters based on samples collected during the period July–October 1983 (Hunt, 1985a). The distribution within the Irish Sea is consistent with the residual advective water movements deduced from general hydrographic observations (Howarth, 1984). In simple terms, North

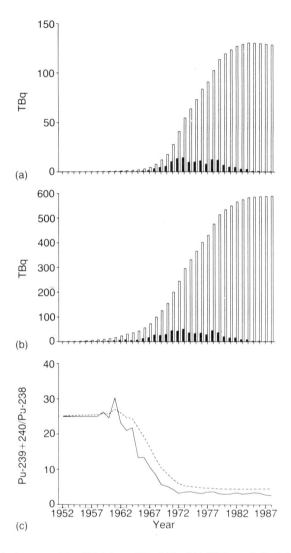

Figure 9.2 Discharges of Pu-238 (*a*), and Pu-239+240 (*b*) from Sellafield to the north-east Irish Sea; the filled bars denote the annual discharge and the open bars denote the decay-corrected cumulative environmental inventory. (*c*) The Pu-239+240:Pu-238 ratio in the effluent (full line) and the decay-corrected cumulative environmental inventories (dashed line).

Atlantic water enters through St George's Channel in the south and the main flow passes northward to the west of the Isle of Man. A minor component of the flow enters the eastern Irish Sea to the north of Anglesey and moves anti-clockwise round the Isle of Man before rejoining the main flow to exit through the North Channel. Dispersion of the caesium to the west of the Isle of Man and into the southern Irish Sea is due to turbulent diffusion driven by the tides and the non-uniform influence of varying wind strength and direction. The distribution of the nuclide also shows some evidence for southerly drifts along the coasts of Ireland, Wales and Cumbria. After leaving the Irish Sea the contamination shows little sign of becoming mixed with North Atlantic water off the continental shelf and moves northward through the Hebridean Sea and around Scotland into the North Sea. The caesium then moves southward along the Scottish and English coasts but is gradually mixed eastwards into the main body of the North Sea becoming diluted with relatively uncontaminated North Atlantic water entering mainly from the north between Shetland and Norway, but also from the English Channel (where there is an input of Cs-137 from the French reprocessing plant at Cap de la Hague near Cherbourg). Data from an earlier cruise, in July 1981, show that the Cs then leaves the North Sea northward along the west Norwegian coast to become dispersed in the Norwegian and Barents Sea (see Fig. 9.3*d*).

The corresponding Cs-134 data for the Irish Sea show generally low concentrations with relatively large analytical errors thus reducing the utility of the Cs-137:Cs-134 ratio as a potential indicator of contaminant 'age'. Although the values of the ratio are uniformly low to the east of the Isle of Man (although higher, at 16–25, compared with an average discharge value of 13 for 1983), there is no consistent pattern elsewhere. For the period 1970–78, when the discharges were higher, Jefferies *et al.* (1982) used the change in Cs-137:Cs-134 ratio between the vicinity of the outfall and the North Channel (Stranraer–Larne) to estimate a mean transit time of one year for the contaminant. This estimate was reasonably consistent with the value of 13 months obtained from a serial correlation analysis of the monthly Cs-137 discharges between April 1971 and May 1978 and the mean water concentrations measured in the North Channel. Other authors have also attempted to use changes in the Cs-137:Cs-134 ratio to estimate transit times from the outfall. Baxter *et al.* (1979) deduced a value of 8 ± 3 months for transport to the Firth of Clyde, and Livingston and Bowen (1977) give estimates of 0.7 and 1.6 years (depending on assumptions) for the transit time to the Minch, and also argued for a very rapid subsequent transport to the area of the North Sea between Scotland and Norway. Kautsky (1988), by using data derived from the extraction of caesium from very large water samples and long radiometric counting times, gave estimates of 6–7 years for the transport of Sellafield contamination to Spitzbergen and 7–9 years to the east coast of Greenland. It is evident that the initial assumptions are crucially important and that the estimates are, at best, only a qualitative indicator of the transit times.

The many data which have been collected concerning the widespread dispersion of Cs-137 from the north-east Irish Sea have not only been used for

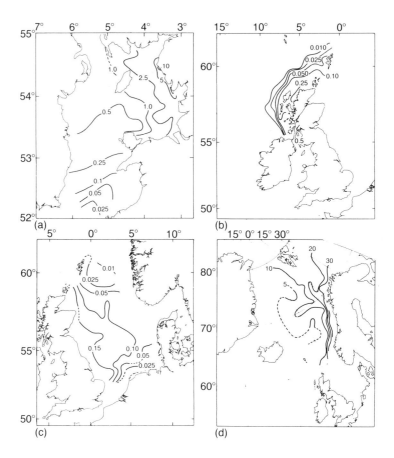

Figure 9.3 Concentrations (Bq kg^{-1}) of Cs-137 in filtered surface sea water from: (*a*) the Irish Sea, September 1983; (*b*) the north-west of Scotland, October 1983; (*c*) the North Sea, August–September 1983; (*d*) the Norwegian and Barents Sea, July 1981.

radiological assessment purposes. Clearly, the distributions largely represent the consequences of water transport and, thus, have been used both as a basis for the development of water circulation models (Jefferies *et al.*, 1982; Hallstadius *et al.*, 1987; Jefferies and Steele, 1989) and as a means of validating models developed from the principles of fluid dynamics (Prandle, 1984; Djenidi *et al.*, 1988).

9.3.2 Cs-137 in the sediments of the eastern Irish Sea

Although the K_D for Cs is relatively low, the historically high discharge rates to the north-east Irish Sea have resulted in readily measurable concentrations of both Cs radionuclides in the sea bed. Sediment cores were collected in May

1983 from an array of 43 stations off the Cumbria and Lancashire coasts. Each core was sectioned horizontally into 5 cm thick slices at the top, and close to the bottom, of the core; the intervening section was sub-sampled to a depth thickness of no greater than 66 cm separated by 5 cm slices. Thus a fairly typical core yielded samples representative of 0–5, 5–71, 71–76, 76–127 and 127–132 cm depth horizons. Each core segment was sub-sampled vertically at constant horizontal area to reject the outer surface of

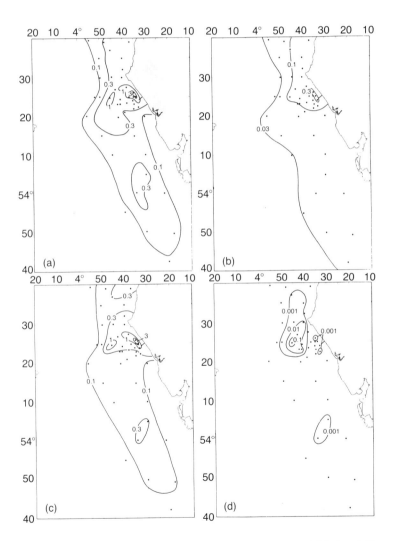

Figure 9.4 Inventory of Cs-137 in the sediments of the north-east Irish Sea (MBq m^{-2}): (*a*) total core; (*b*) surface layer, 0–5 cm; (*c*) subsurface layers, 5–76 cm or the bottom of the core; (*d*) at depths > 71 cm.

the core together with any contamination carried down from the surface layers by the coring action.

Fig. 9.4a shows the distribution of the Cs-137 in terms of the integrated quantity of the nuclide per unit area of the sea bed. There is a pronounced northward displacement of the area of greatest sediment contamination relative to the discharge point; this is a somewhat surprising finding in view of the facts that the bulk of this nuclide is discharged continously and that the distribution in the water shows a southerly drift along the coast. Offshore, the Cs distribution does show a clear southerly set corresponding to the areas of the sea bed consisting of the finest sediments.

In total, the contamination within the 0.1 MBq m^{-2} contour (Fig. 9.4a) amounts to 2.5% of the decay-corrected environmental inventory of Cs-137; this value is confirmed by the data set for Cs-134, and given the difference in half-lives of the two nuclides, the results may be taken as evidence that there have been no major changes in the processes incorporating these materials into the sea bed since discharges began.

The surface layer of the sediment (0–5 cm) accounts, on average for less than 30% of the inventory in the sea bed (Fig. 9.4b), with the greater part of the activity at depths down to 76 cm (Fig. 9.4c). Fig. 9.4d shows those areas where there are detectable traces of the nuclide at depths greater than 71 cm; at one station, in the offshore mud-patch, this amounts to approximately 10% of the total Cs-137 in the core and it is detectable in the sample taken from the 142–147 cm horizon.

The presence of the shorter-lived Cs-134 in association with the Cs-137 provides a potential means of 'ageing' the contamination in the sediment. The data given in Fig. 9.1c show that over the ten years before sampling the Cs-137:Cs-134 ratio for the annual discharge had increased from 4 to 14, and that for the decay-corrected cumulative environmental inventory it had increased from 12 to 40. In isolation from any continuing source the ratio in environmental materials would increase with a doubling time of 2.21 years. Fig. 9.5a shows that there are three offshore areas where the Cs-137:Cs-134 ratio for the total core is greater than 50, i.e. it exceeds the ratio for the decay-corrected cumulative environmental inventory. This may be interpreted to mean that there is a significant delay ($\geqslant 0.7$ years) between discharge and the incorporation of the Cs into the sediments in these areas. Alternatively, in a period of declining discharges (by 70% over the previous five years) and, hence, reduced concentrations in the water column, it may represent reduced incorporation of recent contamination into the sea bed as the overall balance of Cs between water, suspended sediment and seabed sediment is maintained (see below). Fig. 9.5b shows the Cs-137:Cs-134 ratio for surface (0–5 cm) sediments and again there are offshore areas where the Cs appears to be relatively 'old' as it does not reflect either the recent discharge, or the sea water in the region for which a value in the range 16–25 is typical. Because the ratio is greater than that for the cumulative

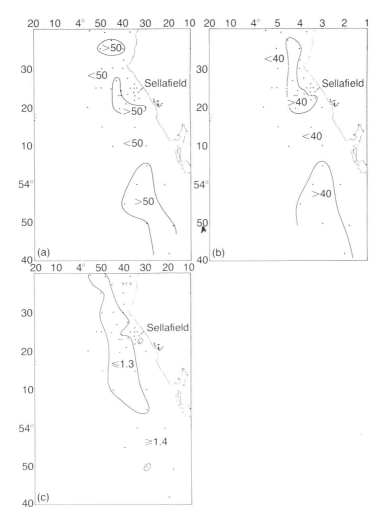

Figure 9.5 Cs-137:Cs-134 ratios: (*a*) total core; (*b*) surface layer, 0–5 cm; (*c*) comparison of ratio at depths greater than 5 cm with that in the surface layer.

environmental inventory, the two alternative explanations offered above remain valid possibilities. The existence of a time lag for the incorporation into the sediment of Liverpool Bay to the south is, perhaps, unexceptionable, but it is rather more surprising for the offshore area immediately to the west and north of the discharge point, where ratios are in the range 40–70, especially as it is completely surrounded by an area in which the ratios (25–39) for surface sediment do appear to reflect an input from recent discharges.

The variation in the Cs-137:Cs-134 ratio with depth in the sediment has been interpreted as an indicator of the degree of mixing of the sediment. Fig. 9.5c shows a comparison of the ratio at depth with that in the surface (0–5 cm) layer; a value of unity shows complete and rapid mixing. The lower values (0.9–1.3), indicative of more rapid mixing (within 1 year) occur mainly in the offshore area.

Table 9.2 Small-scale variability of radionuclide concentrations within a Kasten core

Depth of sub-samples (cm)	Cs-137 (Bq kg^{-1})	$\dfrac{Cs\text{-}137}{Cs\text{-}134}$	Pu-239+240 (Bq kg^{-1})	$\dfrac{Pu\text{-}239+240}{Pu\text{-}238}$
0–5	2.5×10^3	52 ± 3	5.01×10^2	5.0 ± 0.1
	2.2×10^3	48 ± 2	4.94×10^2	4.6 ± 0.2
	2.3×10^3	41 ± 2	4.69×10^2	4.8 ± 0.2
	1.9×10^3	59 ± 4	4.58×10^2	4.6 ± 0.1
30–35	4.8×10^2	45 ± 10	4.06×10^1	6.3 ± 0.2
	3.1×10^2	29 ± 5	7.76×10^1	11.1 ± 0.5
	1.9×10^2	—	1.29×10^1	7.4 ± 0.4
	2.3×10^2	60 ± 30	4.76×10^1	13.4 ± 0.5
60–65	2.1×10^2	—	2.40×10^1	4.7 ± 0.2
	7.1×10^2	60 ± 10	2.36×10^2	4.7 ± 0.1
	1.6×10^2	—	1.54×10^1	5.6 ± 0.3
	5.9×10^2	70 ± 20	1.29×10^2	4.7 ± 0.1
100–105	2.5×10^1	—	4.7×10^{-1}	>20
	ND	—	1.3×10^{-3}	—
	1.8×10^1	—	1.8×10^{-1}	>20
	1.9×10^1	—	2.7×10^{-3}	—
150–155	6.5	—	4.9×10^{-3}	—
	9.5	—	3.2×10^{-2}	—
	4.4×10^1	—	1.58	5.4 ± 0.3
	7.8×10^1	—	9.3×10^{-1}	5.9 ± 0.4

The small-scale variation in the distributions of the nuclides in the sediment is shown by the data summarized in Table 9.2. Four cylindrical (3.8 cm horizontal diameter) subsamples of sediment were taken from the 0–5, 30–35, 60–65, 100–105 and 150–155 cm horizons of a Kasten core (a long box corer with 15 x 15 cm square cross-section) collected from the offshore mud-patch. For the Cs-137, the main points of note are:

1. the concentrations show substantial and increasing variation at each horizon to a depth of 65 cm;
2. the mean concentration at 60–65 cm is greater than that at 30–35 cm;

3. at 150–155 cm there are individual samples with concentrations significantly greater than those at 100–105 cm,

and for the Cs-137:Cs-134 ratio:

4. the ratios at the surface show greater variation than the Cs-137 concentrations, and all are greater than that for the cumulative environmental inventory;
5. there is one sample at 30–35 cm which has a ratio indicative of relatively recent input; and
6. at 60–65 cm, the sample with the highest Cs-137 concentration is likely to be relatively old (4–8 years).

9.3.3 Pu-239+240 in the waters of the Irish Sea

The environmental behaviour of plutonium is likely to be more complex than that of Cs because, in contrast to the presence of the latter in sea water as the simple Cs^+ ion, the element can exist in a number of valence states (Nelson and Lovett, 1978). Hetherington (1976) estimated that at least 96% of the annual input of plutonium was rapidly lost from the water phase and concluded that scavenging by particulate material was the process responsible. Pentreath *et al.* (1984) reported that more than 97% of the plutonium in the Sellafield effluent was present in the reduced (III+IV) valence states, and already almost entirely in particulate form (> 0.22 μm), with the remainder in the oxidized (V+VI) states split evenly between the particulates and solution. Nelson and Lovett (1978) had shown that the K_D values for Pu (III+IV) were of the order of 10^6, whereas those for Pu (V+VI) were less than 10^4 and, therefore, of similar magnitude to that for Cs. Thus, it is not surprising that the majority of the plutonium discharged is rapidly removed to the sea bed with only a few percent remaining in the water column and immediately available for wider dispersion.

The labour-intensive nature of the analytical procedures for the Pu nuclides means that there are rather fewer data for the relatively minor concentrations that remain in the water compartment after discharge. Figs. 9.6a and b show the distributions of Pu-239+240 in filtered sea water, for the Irish Sea (26 stations) and U.K. coastal waters (15 stations), respectively, as derived from surface samples collected in October–November 1985. Comparison of the data for the Irish Sea with that for Cs-137 (Fig. 9.3a) shows that the pattern of dispersion and dilution of the soluble fractions of the two nuclides is qualitatively similar, a result which is consistent with the conclusion above that the plutonium remaining in solution is in the relatively conservative oxidized form. Outside the Irish Sea there are too few data to make detailed comparisons but again the distribution appears qualitatively similar to that for Cs.

In the years before these measurements were done, the Pu-239+240: Pu-238 ratio in the discharge had been in the range 3.0–3.4, while that for the

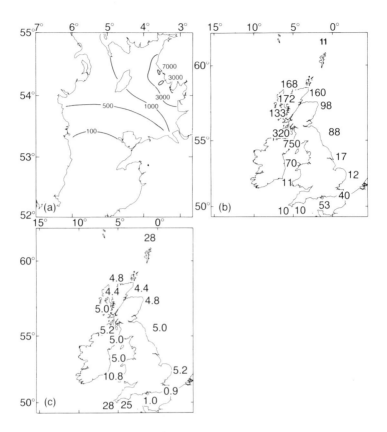

Figure 9.6 Concentrations (μBq 1^{-1}) of Pu-239+240 in filtered surface sea water from: (*a*) the Irish Sea; (*b*) British coastal waters, in October–November 1985; (*c*) Pu-239+240:Pu-238 ratio in filtered surface sea water.

cumulative environmental inventory had stabilized at a value of 4.4 (Fig. 9.2*c*). The limited data available (Pentreath *et al.*, 1984) indicate that there is little variation in the ratio in the different components and streams of the effluent. Within the Irish Sea (Fig. 9.6*c*), the distribution of the ratio, with values in the range 4.0–5.2, shows no consistent pattern, but it is evident that the soluble fraction of the plutonium is not entirely derived from recent inputs. Indeed, it may be concluded that complete equilibrium has been attained between the soluble phase and the cumulative environmental inventory, most of which resides in the sea bed. These results imply that, as discharges decline, the sea bed will become a net source of plutonium to the water column.

Outside the Irish Sea the Pu-239+240:Pu-238 ratio does show variation (Fig. 9.6*c*). To the north of the Shetland Isles and to the south and west of

Cornwall the isotope ratio (25–28) is typical of fallout from weapon testing and there is little, if any, contribution of Sellafield plutonium to these samples. At the southern entrance to the Irish Sea the isotope ratio is intermediate between those for fallout and the Sellafield discharges and is clear evidence for the mixing of incoming North Atlantic water with the contaminated water of the Irish Sea. In the central and eastern regions of the English Channel the ratio is around unity and shows clearly the influence of the input from the French reprocessing plant at Cap de la Hague near Cherbourg. In Scottish coastal waters and the eastern North Sea the plutonium is very largely of Sellafield origin and the data support the pattern of water movement deduced from the Cs distributions. Pu-239+240:Pu-238 data (unpublished) derived from samples collected in July 1985 from the Norwegian and Barents Seas show that the plutonium originating from Sellafield moves northward close inshore along the Norwegian coast (see Fig. 9.3*d* for Cs data) and into the Barents Sea. There is some mixing offshore and in the western Barents Sea as evidenced by isotope ratios intermediate between that for the Sellafield discharge and that for fallout.

9.3.4 Plutonium in the sediments of the eastern Irish Sea

The distribution of plutonium-239+240 in the sea bed is shown in Figs. 9.7*a–c*. For the total Pu activity in the sediment, the pattern generally follows the distribution of fine sediment, as expected, and is very similar to that for the Cs-137 (Fig. 9.4*a*). Approximately 65% of the total inventory has been mixed to depths greater than 5 cm and again, mainly in the offshore areas, there are traces of the nuclides at depths greater than 71 cm (Fig. 9.7*c*) with one station showing a significant concentration at 142–147 cm. Within the area of the 0.01 MBq m^{-2} contour the Pu-239+240 content amounts to 30% of the total decay-corrected environmental inventory. When these data on environmental concentrations first became available the Pu-238 content of this area was estimated to be almost 90% of the environmental inventory based on the discharge data published by Stather *et al.* (1986). On the reasonable assumption that the environmental behaviour of the Pu-238 and the Pu-239+240 would be very similar, the magnitude of the difference between the results for the nuclides indicated a substantial deficiency in the data for the discharges, as was known from earlier work (Hetherington, 1976; 1978). By using the latest estimates of the inputs of the Pu nuclides based on the data from the Maryport Harbour core (Kershaw *et al.*, 1990) the Pu-238 content has been reassessed at 29% of the decay-corrected environmental inventory. This improved, and remarkable, agreement between the data for the different Pu nuclides provides substantial independent support for the reassessed historical inputs of Pu-238 and Pu-239+240.

The Pu-239+240:Pu-238 ratio in sediment has been used on a number of occasions to attempt to understand the processes of sediment accumulation

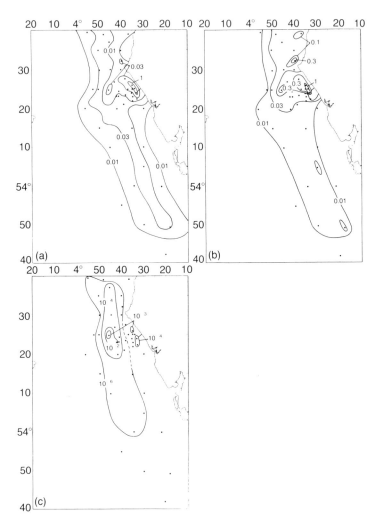

Figure 9.7 Inventory of Pu-239+240 in the sediments of the north-east Irish Sea (MBq m^{-2}): (*a*) total core; (*b*) at depths >5 cm; (*c*) at depths >71 cm.

and movement in this area of the sea bed (Hetherington, 1976; 1978). The most recent studies (Kirby *et al.*, 1983) have concluded that there is no evidence for substantial accretion of fine sediment in the area, and that the various profiles of Pu-239+240:Pu-238 ratio observed in sediment cores represent 'snap-shot' pictures of a continuing, and as yet incomplete, mixing process largely driven by the populations of benthic in-fauna. The large echiuroid worm *Maxmulleria lankesteri* has been shown to be capable of redistributing activity deposited at the surface of the sediment to depths down to 35 cm (Kershaw *et al.*, 1983, 1985) and there is circumstantial evidence implicating the small crustacean *Callianassa subterranea* as a potential generator of additional pathways for deeper penetration of the radionuclides into the sea bed (Swift, in preparation).

The data presented in Table 9.2 for the plutonium nuclides show more clearly than the corresponding data for caesium the small-scale variations in the distributions of the nuclides with time. The surface layer appears to be relatively well-mixed, there being less variation for the Pu-239+240 than is the case for Cs-137. The Pu isotope ratio is similar to that of the Pu in solution in the water column and represents closely the cumulative inventory. At 30–35 cm the concentrations have declined to approximately 10% of, and show substantially greater variation than, those at the surface. The isotope ratio values are susceptible to a number of interpretations as regards the time of origin:

1. if they are representative of material discharged in a given year, then the relevant period is 1965–68;
2. if they are representative of the cumulative discharge then the period is 1968–71;
3. they could represent a discontinuous mixture of pre-1965 material (high ratio values) with that discharged more recently (low ratio values).

Whichever interpretation is placed on the data, each shows that it is possible for the contaminant to be incorporated into the sediment at depth and then remain effectively undisturbed for substantial periods of time. At 60–65 cm the overall variation in concentration again shows an increase but individual sample concentrations are higher than those at 30–35 cm with the result that the average concentration is increased to approximately 20% of that at the surface. However the isotope ratio for these samples is interpreted, it is consistent in indicating that a substantial quantity of plutonium of more recent origin than that at 30–35 cm has been incorporated into the sediment. The concentrations at 100–105 cm show a substantial decrease but with increased variation. For the two samples for which a value could be determined, the plutonium isotope ratio indicated that the material had been discharged prior to the mid-1960s. The deepest samples from 150–155 cm show slightly increased concentration and similar variability relative to those at 100–105 cm, but the isotope ratios imply a probable origin in the period 1970–75 unless process 3 above is operative. These results are not atypical and show that the plutonium is very far from being at an equilibrium distribution in the environment.

9.3.5 The Cs-137:Pu-239+240 ratio

The relative behaviour of the Cs-137 and Pu-239+240 can be seen rather more clearly in terms of the isotope activity ratio. In the ten years prior to the sediment sampling the ratio in the effluent had shown wide fluctuations and the decay-corrected cumulative environmental value had increased to a value of 59 (see Fig. 9.8*a*). The isotope ratio values for surface sediment (0–5 cm) and the total core are given in Fig. 9.8*b*. The values are less than that either for the recent discharges or the cumulative environmental inventory and increase with distance from the outfall. Both these results confirm the more soluble, less particle-reactive nature of the Cs-137 compared with Pu-239+240 and show quite clearly the more efficient scavenging of the latter from the water column. For the surface sediment, equivalent values of the ratio generally occur at greater distances from the discharge point than for the total core. If it is assumed that the surface sediment is more likely to reflect recent discharges than the total core, this observation might be taken to be at variance with the fact that the ratio in the discharge had tended to increase. However, it must be remembered that the total discharges had been declining (Figs. 9.1 and 9.2), and it appears quite possible that the sediment had become, at the time of sampling, a source of Cs to the water column. The presence of offshore areas of sediment where the Cs contamination is apparently 'older' (in terms of the Cs-137:Cs-134 ratio) than the cumulative environmental inventory (see Fig. 9.5 and §9.3.2) might also be partly explained by this release process.

9.4 CONCLUSIONS

The dispersive capacity of the sea is frequently employed to dilute contaminant inputs to concentrations that are judged to be harmless. Once a waste material has been discharged, or dumped, into coastal waters, the subsequent dispersal, although perhaps predictable, is essentially uncontrollable. Thus measures to reduce environmental impact can only be effected before disposal.

The currently accepted radiation protection philosophy recommends that, within overall dose limits, all exposures shall be kept as low as reasonably achievable (ALARA), economic and social factors being taken into account (ICRP, 1977). Although the radiation exposure of the general public arising from liquid waste discharges in coastal waters has been maintained within prescribed limits (Hunt, 1985*a*), there has been continuing pressure for reductions in the degree of human exposure considered to be acceptable. Together with the application of the ALARA principle, this has led to reductions in the quantities of radionuclides discharged to sea and consequent decreases in the concentrations of the radionuclides dispersed in the environment. This trend seems likely to continue.

The presence of the artificial radionuclides in the sea has provided a spur for the study of the processes that govern their distributions, and, incidentally, those of other contaminants. In consequence, there has been a substantial improvement in the understanding of the marine system both directly,

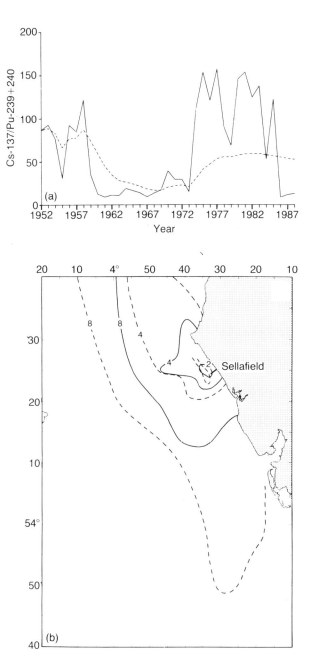

Figure 9.8 (*a*) Cs-137:Pu-239+240 ratio in the effluent (full line) and the decay-corrected cumulative environmental inventories (dashed line); (*b*) Cs-137:Pu-239+240 ratio in sediment samples: full lines, total core; dashed lines, surface layer, 0–5 cm.

and in the provision of supporting evidence for data derived by independent means. The results obtained from studies of the dispersion of artificial radio-nuclides from the Sellafield outfall permit a number of general conclusions to be drawn:

1. the behaviour of caesium as the Cs^+ ion indicates that soluble materials become widely dispersed from the input, but a significant fraction may be held back by interactions with local fine sediment;
2. the behaviour of plutonium indicates that the greater part of particle-reactive materials will be confined to fine sediments in the vicinity of the discharge point, but changes in chemical speciation can result in a small fraction becoming widely dispersed;
3. within the sediment compartment an equilibrium distribution of the contaminant may take many decades to become established;
4. if inputs to the environment decline, it is likely that the sediment inventory of a contaminant will become a secondary source to the water column.

Thus it may be inferred that almost any waste material discharged into the coastal marine environment is likely to be present in the sea, although not necessarily detectable, at great distances from the source; the rate and extent of this dispersal will be dependent on physical, chemical and biological processes.

As inputs of the artificial radionuclides decline with the imposition of more restrictive discharge authorizations, there will be fewer possibilities to use them as tracers of marine processes. It should not be forgotten, however, that the distributions of the present environmental inventories will continue to evolve and provide opportunities for the study of dispersive processes. In addition, there is also present in the marine environment a suite of natural radionuclides which exhibit a wide range of geochemical properties. These can be, and are, utilized in appropriate circumstances to yield similar information concerning the rates and magnitudes of marine processes as a function of space and time.

ACKNOWLEDGEMENTS

It is a pleasure to acknowledge the many contributions made by our colleagues within the Aquatic Environment Protection Division of the Directorate of Fisheries Research to the work discussed in this paper.

REFERENCES

Baxter, M.S., McKinley, I.G., MacKenzie, A.B. and Jack, W. (1979) Windscale radiocaesium in the Clyde Sea area. *Mar. Pollut. Bull.* **10**, 116–120.
BNFL (1984) *Annual report on radioactive discharges and monitoring of the environment 1983*. British Nuclear Fuels plc, Warrington, Cheshire, U.K.

BNFL (1985) *Annual report on radioactive discharges and monitoring of the environment 1984.* British Nuclear Fuels plc, Warrington, Cheshire, U.K.

BNFL (1986) *Annual report on radioactive discharges and monitoring of the environment 1985.* British Nuclear Fuels plc, Warrington, Cheshire, U.K.

BNFL (1987) *Annual report on radioactive discharges and monitoring of the environment 1986.* British Nuclear Fuels plc, Warrington, Cheshire, U.K.

BNFL (1988) *Radioactive discharges and monitoring of the environment 1987.* British Nuclear Fuels plc, Warrington, Cheshire, U.K.

BNFL (1989) *Radioactive discharges and monitoring of the environment 1988.* British Nuclear Fuels plc, Warrington, Cheshire, U.K.

Djenidi, S., Nihoul, J.C.J. and Garnier, A. (1988) Modele mathematique du transport des radionuclides sur le plateau continental nord-europeen. In *Radionuclides: a tool for oceanography* (ed. J.-C. Guary, P. Guegueniat and R.J. Pentreath), pp. 373–383. London: Elsevier Applied Science.

Hallstadius, L., Garcia-Montano, E., Nilsson, U. and Boelskifte, S. (1987) An improved and validated dispersion model for the North Sea and adjacent waters. *J. Environ. Radioactivity* **5**, 261–274.

Hetherington, J.A. (1976) The behaviour of plutonium nuclides in the Irish Sea. In *Environmental toxicity of aquatic radionuclides: models and mechanisms* (ed. M.W. Miller and J.N. Stannard), pp. 81–106. Michigan: Ann Arbor Science Publishers.

Hetherington, J.A. (1978) Uptake of plutonium nuclides by marine sediments. *Mar. Sci. Commun.* **4**, 239–274.

Howarth, M.J. (1984) Currents in the eastern Irish Sea. *Oceanogr. Mar. Biol. Ann. Rev.* **22**, 11–53.

Hunt, G.J. (1985*a*) Radioactivity in surface and coastal waters of the British Isles, 1983. *Aquat. Environ. Monit. Rep.* MAFF Directorate, Fisheries Research, Lowestoft, **12**, 1–46.

Hunt, G.J. (1985*b*) Timescales for dilution and dispersion of transuranics in the Irish Sea near Sellafield. *Sci. Total Environ.* 46, 261–278.

IAEA (1985) Sediment K_ds and concentration factors for radionuclides in the marine environment. *Tech. Rep. Ser.* No. 247 (73 pages.) Vienna: IAEA.

ICRP (1977) Recommendations of the International Commission on Radiological Protection. *Ann. ICRP* **1**, 1–53.

Jefferies, D.F., Steele, A.K. and Preston, A. (1982) Further studies on the distribution of Cs-137 in British coastal waters. *Deep-Sea Res.* **29**, 713–738.

Jefferies, D.F. and Steele, A.K. (1989) Observed and predicted concentrations of caesium-137 in seawater of the Irish Sea 1970–1985. *J. Environ. Radioactivity* **10**, 173–189.

Kautsky, H. (1988) Determination of distribution processes, transport routes and transport times in the North Sea and northern North Atlantic using artificial radionuclides as tracers. In *Radionuclides: a tool for oceanography* (ed. J.-C. Guary, P. Guegueniat and R.J. Pentreath), pp. 271–280. London: Elsevier Applied Science.

Kershaw, P.J., Swift, D.J., Pentreath, R.J. and Lovett, M.B. (1983) Plutonium redistribution by biological activity in Irish Sea sediments. *Nature, Lond.* **306**, 774–775.

Kershaw, P.J., Swift, D.J., Pentreath, R.J. and Lovett, M.B. (1985) The incorporation of plutonium, americium and curium into the Irish seabed by biological activity. *Sci. Total Environ.* **40**, 61–81.

Kershaw, P.J., Woodhead, D.S., Malcolm, S.J., Allington, D.J. and Lovett, M.B. (1990) A sediment history of Sellafield discharges. *J. Environ. Radioactivity.*

Kirby, R., Parker, W.R., Pentreath, R.J. and Lovett, M.B. (1983) Sedimentation studies relevant to low-level radioactive effluent dispersal in the Irish Sea (66 pages). Wormley: IOS Report No. 178.

Livingston, H.D. and Bowen, V.T. (1977) Windscale effluents in the waters and sediments of the Minch. *Nature, Lond.* **269**, 586–588.

Nelson, D.M. and Lovett, M.B. (1978) Oxidation state of plutonium in the Irish Sea. *Nature, Lond.* **276**, 599–601.

Pentreath, R.J., Lovett, M.B., Jefferies, D.F., Woodhead, D.S., Talbot, J.W. and Mitchell, N.T. (1984) Impact on public radiation exposure of transuranium nuclides discharged in liquid wastes from fuel element reprocessing at Sellafield, United Kingdom. In *Radioactive waste management*, vol. 5, pp. 315–329. Vienna: IAEA.

Prandle, D. (1984) A modelling study of the mixing of Cs-137 in the seas of the European continental shelf. *Phil. Trans. R. Soc. Lond.* A **310**, 407–436.

Stather, J.W., Dionian, J., Brown, J., Fell, T.P. and Muirhead, C.R. (1986) *The risks of leukaemia and other cancers in Seascale from radiation exposure* (157 pages). Chilton: NRPB-R171 Addendum, NRPB.

Part Four

Degradation

Chairman's introduction

A.T. Bull

Until relatively recent times environmental pollution has been pre-occupied with the consequences of past and present mismanagement, i.e. with symptoms, rather than with the causes of pollution. Such a focus has influenced thinking on the means of pollution control, and has lead to an emphasis on treatment technologies and a comparative neglect of process developments or modifications that minimize pollution.

The principal points at which pollution control can be exercised are as follows:

1. in-process treatment;
2. end-of pipe treatment;
3. remediation of polluted sites;
4. modification of existing processes;
5. introduction of new processes and products.

The theme of the first three options is clean-up; either-in-house, during the manufacturing process itself, to remove a target contaminant from the product or the final discharge stream, or at the point of discharge of the wastes; or *in situ* as a result of historical pollution of soil and water.

The remaining two options for pollution control are founded on long-term strategies in which minimization and recycling are clearly preferable to treatment and disposal both in social and commercial terms. Modification of processes to alleviate pollution loads take a variety of forms: improved catalytic efficiency, energy and materials recycling, and process integration are examples of this philosophy. They are all manifestations of what Frosch and Gallopoulos (Chapter 16) call the management of 'industrial eco-systems'. Hallstead (Chapter 14) describes a good case of materials recycling in flue gas desulphurization to produce gypsum. Among the pointers to new process and product development may be cited those that eliminate the use of organic solvents, substituting the use of chlorine in pulp bleaching, or in reducing the demand for water. For example, Bayer's thermoplastic Novodur is now produced by a water-free polymerization with annual savings of 400 000 m^3 of water and the elimination of 800 t of waste.

In relation to this, then, degradation is not an ideal option. In effect, it represents a waste of materials, unless some of the products can themselves be brought into use. It is most valuable where a hazardous waste is involved which must be eliminated effectively and permanently. But it can be equally valuable where large quantities of a non-hazardous waste are being produced which cannot be eliminated by other means. These two considerations form the basis of the two papers, by Bull and by Anderson and Pescod, in this section.

It is possible that at least some of the products of degradation can be used to advantage. Incineration is an important method of degradation where important technical developments are being made. In the new incineration system installed by Bayer at Leverkusen, toxic substances of high calorific value are burnt in hazard waste incinerators and the heat generated is used to dry and incinerate sewage sludge from a waste-water treatment plant, thereby saving on landfill disposal. Burnt flue gases are thermally oxidized and the heat is transformed into pressurized process steam. This integrated operation enhances the effectiveness of waste treatment and produces energy savings of about 10% of the original requirements. The benefits of specific energy audits in the handling of wastes, as well as the general waste audits discussed by Backman and Lindhqvist (Chapter 2), are very clear.

The two papers emphasize that degradation is not an easy option. Whatever processes are employed, they are sensitive to operating conditions and the presence of other substances in the waste. In the biological degradation systems the operator is effectively in charge of a whole, complex and potentially sensitive, ecosystem. The required outcome is that hazardous or troublesome wastes are reduced to levels where they no longer cause a problem. In all degradation treatment systems, it is therefore important to maintain appropriate monitoring to ensure that this outcome is being achieved.

10

Degradation of hazardous wastes

A.T. Bull

10.1 INTRODUCTION: THE CONSEQUENCES OF MISMANAGEMENT

The threat to public health resulting from the disposal of hazardous chemical wastes in the environment has been well known and highlighted by the notorious histories of Love Canal, the Valley of the Drums, and the village of Lekkerkerk. However, the extent to which contaminated land poses a direct environmental hazard or an indirect hazard through polluted groundwater is only now being fully appreciated. Estimates of the size of the problem vary considerably. In the United States 20 000 hazardous waste sites have been identified, 1500 of which have been assigned to the National Priorities List. Whilst the Commission of the European Communities estimates that over 50 000 sites exist in the EC (Gieseler, 1987), other inventories suggest that *ca.* 100 000 polluted sites occur in The Netherlands alone of which 4000 require urgent remediation (De Brauw, 1989). The only certainties in this context are that the problem is immense and still growing, and that comprehensive information on the types of pollution is largely lacking. Treatment technologies for hazardous wastes require to be developed to cope with these historical problems but also to prevent further pollution. Options for dealing with the latter situation are considered in the next section.

The question of what constitutes a harmful or hazardous waste is germane to this discussion. In the U.K. the Environment Protection Bill defines harm as harm to the health of living organisms and, in the case of man, includes offence to any of his senses or harm to his property. Others (see Kramer, Chapter 3) argue that all wastes should be deemed hazardous, with

The Treatment and Handling of Wastes
Edited by A.D. Bradshaw, Sir Richard Southwood and Sir Frederick Warner
Published in 1992 by Chapman & Hall, London, for The Royal Society
UK ISBN 0 412 39390 5, USA ISBN 0 442 31461 2

derogation being granted only for demonstrated harmless wastes, and that harm should also embrace abiotic effects and aesthetics. Similarly, how should clean sites be distinguished from polluted sites? In the U.K. the current Guidance Note the Inter-departmental Committee on the Re-development of Contaminated Land infers that any hazard posed by an individual site will be assessed against the proposed use of that site. Pollutants are categorized as: (1) those presenting a hazard even in very low concentrations; (2) those producing a measurable effect on a target at a given concentration, and (3) those for which no dose–effect relationships have been established. To determine the significance of the contamination, trigger concentrations have been introduced which relate to the intended use of the site. However, there remains a dearth of information on the degradation and the environmental effects of mixtures of contaminants (see Bull, 1980) and for which guidelines on specified contaminants will have only partial relevance. It may be prudent, therefore, to adopt tests which assess mutagenic or carcinogenic potential, such as the Ames test, for screening polluted sites; evaluations based on such screening are now being reported (Jones & Peace, 1989; Lloyd, 1989). Readers wishing to pursue the issue of environmental quality objectives and standards, and of the control of 'Red List' substances are referred to the article of Agg and Zabel (1989).

Most commentaries on hazardous wastes and the environment are set in the context of sustainable development for fulfilling the 'needs of this generation without depriving future generations of the possibility of using the environment to meet their own needs' (World Commission on Environment and Development, 1987). Consequently research and technology development priorities need to focus not only on the clean-up of wastes (historical and current) but also on the amelioration of extant processes and the introduction of new, environment-compatible processes and products.

However, given the state of environmental pollution as it exists now, we will examine the technological options which are available, or in development, to address clean-up. In this, degradation is an important option. It allows us to reduce amounts of hazardous wastes, either left by previous mismanagement or still being produced by current inefficient processes, to levels at which they no longer pose a threat to our environment. But as an approach it must be placed in perspective with all the other different treatment technologies that are available.

10.2 THE TECHNOLOGIES

The most obvious fact about clean-up technologies is the bewildering choice which is available; it is more accurate to say apparently available because relatively few of them have been proven and are in routine use. These technologies can be grouped on the basis of the scientific and engineering principles involved, or, on the basis of how and where the technology is

Table 10.1 Treatment technologies for hazardous wastes

Physical
Soil washing
Air (or steam) stripping
Carbon adsorption
Ion exchange
Solidification
Vitrification

Thermal
Rotary kiln incineration
Infra-red incineration
Plasma arc incineration

Chemical
Solvent extraction
Oxidation (photolytic; electrolytic; chemical)
Ozonolysis
Electrokinetic removal

Biological
Land farming
Composting
Bioreactor processes (aerobic; anaerobic; biofilm)
Bioenrichment
Bioaugmentation
Landfill

---

implemented (i.e., containment on site; treatment *in situ*; treatment on site; treatment or disposal off-site). Some indication of the range of physical, thermal, chemical and biological treatment technologies is given in Table 10.1. Although there have been several reports on the comparative efficacy of treatment technologies (Table 10.2 provides an example of such predictions), ultimately detailed case-by-case analyses must be made.

In 1989 the U.S. Environmental Protection Agency published a broad assessment of international technologies (particularly those from Europe, Canada and Japan), that might find application for the remediation of Superfund sites. The following criteria were used in the assessment: (1) function: object and applicability of the technology; (2) description: details of operating principles and design features; (3) performance: demonstrable clean-up on waste sites; (4) limitations: features that limit applicability; (5) economics, and (6) status: of research, development and availability. The majority of the technologies (63/95) were recommended for continued monitoring (Nunno *et al.*, 1989). The U.S. EPA also established the Superfund Innovative Technology (site) Program. By 1988 29 U.S. companies were contributing to a demonstration programme (representing

Table 10.2 Predicted effectiveness of treatment technologies in decontaminating soil

Organic contaminant	*Technology Thermal*	*Solvent Extraction*	*Dechlorination*	*Bioremediation*
Polynuclear aromatics	D	P	N	D
PCBs, dioxins, furans	D	P	P	P
Halogenated aromatics	D	P	P	P
Halogenated aliphatics	D	P	P	D
Heterocyclics	D	D	N	D
Nitro-compounds	D	P	N	D
Polar, non-halogenated compounds	D	P	N	D

D, demonstrated effectiveness; P, potentially effective; N, no expected effectiveness. (Information from Superfund 88.)

a full spectrum of technologies), seven had been selected for funding under an emerging technologies programme (metals removal; laser photochemical oxidation; solvent soil washing; electroacoustic soil decontamination; and *in situ* oil removal) while a further five technologies were selected for development through pilot scale to field evaluation by the EPA before seeking commercial partners (mobile soil washer; dehalogenation by polyethylene glycolates; regeneration of spent activated carbon; and electrokinetic removal of contaminants) (U.S. EPA, 1988; Hill, 1988).

Two final points should be made at this juncture. First, the choice of technologies for treating or disposing of hazardous wastes is heavily influenced by political and public perceptions, by geomorphological conditions, and by cost. For example, in the U.K. landfill is the prime method for hazardous waste disposal, but in other EC countries the practice is severely restricted or banned as a result of secondary contamination of soils and groundwaters (Table 10.3). In contrast, thermal processes have been developed extensively in other parts of Europe but constitute a minor disposal route in the U.K. (less than 10% of municipal solid waste). Moreover, the public outcry against incineration in the North Sea has forced larger incinerators to be built on land. Secondly, although physicochemical and thermal processes dominate hazardous waste treatment at present, bioremediation technologies can be economically competitive and, correspondingly, are now receiving increased attention and commercialization. Table 10.4 assembles some recently gathered cost comparisons.

10.3 THE IMPACT OF BIOTECHNOLOGIES

Biotechnological contributions to hazardous waste management relate to the development of biological catalysts for the degradation, detoxification

Table 10.3 Hazardous waste disposal route in the U.K.

Method	Percentage of total arisings[a]
Landfill	78.7
Solidification	3.6
Mineshafts	1.6
Sea disposal	7.0
Chemical treatment	7.5
Incineration	1.6

[a] 4.8 Mt per year.
Data taken from The Management of Hazardous Waste in the United Kingdom, April, 1989. Her Majesty's Inspectorate of Pollution, London.

or accumulation of toxic chemicals and metals. The rather slow and sometimes inconsistent introduction of biotechnological processes has, in part, reflected early, over-optimistic claims in some quarters for the clean-up of hazardous wastes and contaminated sites following the application of specialized or customized, microbial mixes. Failure of such concoctions frequently led to a return to proven physical chemical or thermal processes which could deliver 'nine nines' efficiency. During the past decade biotreatment processes have been designed and successfully implemented as a result of applying scientific and engineering principles, and being fully cognizant of the particular waste problem to be resolved. The starting point for considering a biotechnological solution must be the precise definition of the pollution problem; this obvious point of departure is often under-researched. Previously we have suggested that a check list including the following issues should be addressed (Bull *et al.*, 1988).

1. Analysis of the problem: is it a single component of the waste that is difficult to treat, or is it a multicomponent problem?
2. Can the manufacturing plant (or process) be adapted to remove the problem component or source?
3. Is the waste amenable to treatment by application of specialized microorganisms, or by specific enzymes?
4. Is biorecovery of materials (or energy) from the waste an option?
5. Does the pollution arise as a result of (i) deliberate discharge; (ii) a side effect of application (e.g. pesticides); (iii) mining, dumped container, or landfill leachate, or, (iv) accidental spillage?

The environment in which biotreatment of hazardous wastes has to be affected requires further comment; the wide variety of environments dictates that a comparable variety of treatment strategies has to be deployed. Thus, conventional bioreactor technology by using micro-organisms, or enzymes, will be appropriate for treating well-defined industrial waste streams and contaminated products (e.g. *Pseudomonas* species or the specific hydrolase

Table 10.4 Approximate unit costs for remedial technologies

Technology	Cost range (£/cu.m)
Landfill (1) U.K.	25–120
(2) U.S.A.	100–200
Thermal (1) Holland, off site	25–100
(2) U.S.A., off site	100–500
(3) U.S.A., on site	75–300
Air stripping of volatiles	20–50
Soil washing	35–100
Bioremediation	5–75

Data kindly provided by Dr R. McLean, Arthur D. Little, Cambridge.

therefrom for detoxifying parathion). Communities of micro-organisms rather than mono-species populations may be required to treat more complex wastes or wastes containing recalcitrant chemicals (see Bull, 1989; Slater and Bull, 1982). And finally, in circumstances where a natural environment is polluted by an often ill-defined mixture of hazardous waste an ecosystem approach has to be made to clean-up of the site.

Fundamental studies of the biodegradation of the major organic chemical pollutants are well advanced, as are the means of isolating and selecting an organism (or community of organisms) having appropriate catabolic activity. However, in the context of waste treatment, it will not be sufficient simply to select organisms solely on their ability to degrade target chemicals; selection must be made with reference to the design and operation of the treatment process, and the constancy, or otherwise, of the composition of the waste stream(s). Design of cost-effective biotreatment processes will take account of the following criteria:

1. the overall rate of detoxification or complete mineralization and the kinetic order of reaction;
2. the target final concentration required, and the capacity for reducing it further in the face of changing legislation;
3. the physiology of the micro-organism(s). Among the prime considerations will be: safety; ease of handling (process start-up; process maintenance including requirements for nutrient additions, pH and dissolved oxygen; reactor half-life); resistance to waste stream components other than the target pollutant (acidity, alkalinity, salinity, heavy metal ions);
4. competitiveness of the micro-organism(s) against contaminants (bioreactor processes) and indigenous microflora (*in situ* bioremediation operations).

Although it is impracticable to develop universal biocatalysts for the treatment of all hazardous chemicals it is realistic to develop reactor configurations that can treat multi-component wastes whose composition may oscillate with time. Waste treatment bioreactors are usually based on a small number of conventional types, of which stirred tank, activated sludge, anaerobic and biofilm systems predominate (see Anderson and Pescod, Chapter 11, for descriptions of some of these reactors). Biofilm, or filter, reactors have proved particularly useful for treating hazardous wastes. Among their attributes is a superior capacity to accommodate shock loadings without major breakthrough of the pollutant materials into the discharge stream and to treat highly dilute pollutants. Typical operating costs for submerged biofilm reactors range from $0.05 to $3.0 per 1000 gal; average costs for carbon adsorption treatment (which does not eliminate the pollutant) may be 10 to 400 times greater. Nevertheless, the criticism that biotechnology in hazardous waste treatment is 'an unfulfilled promise' (Nicholas, 1987) warrants our serious evaluation lest previous disenchantment with this approach is to reappear. Examples of the bioremediation of polluted sites will serve to advertise the efficacy of biotechnology. This does not imply that effective applications of biotechnology in degrading hazardous chemicals in both end-of-pipe discharges and product streams have not been made, but rather that details of some of the best examples are subject to commercial secrecy.

10.4 BIOREMEDIATION OF POLLUTED SITES

Bioremediation is a term that encompasses biological methods for the clean-up of contaminated land and water. It can imply the complete restoration of a site so that its original multifunctional use is recovered, or, reclamation of a site for a particular intended use. Clearly the degree of pollution removal will vary in these two cases. Bioremediation operations may be made either on-site or off-site, *in situ* or *ex situ*. There is growing interest in *in situ* remediation operations where treatment occurs at the original location and does not incur excavation. However, irrespective of the type of operation, bioremediation involves the deployment of micro-organisms to detoxify or mineralize hazardous chemicals. Such chemicals are utilized as sources of nutrients and/or energy by micro-organisms or are degraded by means of co-metabolic transformations. The rate and extent of the degradation are functions of the prevailing physicochemical conditions in the soil or water, and bioenrichment methods, whereby the site is supplemented with nutrients, pH modifiers or oxygen, are often used to enhance the microbial activity. The natural selective pressure of polluting chemicals at a contaminated site almost certainly will have led to the development of indigenous microbial populations that can metabolize them (assuming that such chemicals are biodegradable). But again, the seeding of such sites with populations of competent micro-organisms produced in fermenters is used to speed up the bioremediation; this approach is known as bioaugmentation.

The first attempts at bioremediation *in situ* were based on the principal of land farming. Contaminated soil, spread in relatively thin layers, is supplemented with nutrients and micro-organisms, has its pH adjusted if necessary and is tilled periodically to aerate. The technique is demanding in terms of land area and remediation is slow; Lapinskas (1989) cites 2 years to achieve *ca.* 90% reduction in a fuel oil contaminated land farming project. Bioremediation rates can be increased significantly in engineered soil banks. Here banks of contaminated soil are heaped onto high density plastic liners such that run off leachate can be collected. The latter may be recycled as an irrigant for the bank or treated separately in a bioreactor. The treatment time to degrade hydrocarbon pollutants is often halved in comparison to conventional land farming. A logical extension of the above bioremediation approaches is the pump-and-treat type of process; this option may be desirable in circumstances where the contaminated soil is inaccessible (e.g. close to or beneath buildings) or where groundwater is required to be treated. Soil leachates, or groundwaters, are pumped to the surface, treated under controlled conditions in a bioreactor, and the cleaned water returned to the ground. The clean-up of soils and sub-soils by this means is totally dependent on their permeabilities. Moreover, the polluting chemicals when dispersed in soil and groundwork are present at extreme dilutions. Thus, *in situ* bioremediation presents immense challenges, and opportunities, for the biotechnologist and necessitates the closest cooperation with engineers and hydrologists for achieving success.

10.5 CASE HISTORIES

The first case to be discussed concerns the *in situ* bioremediation of an urban, industrial site in the U.K., the former Greenbank Gas Works at Blackburn. Coal gasification occurred at these Works for *ca.* 70 years and over half of the 10 ha site was contaminated with biodegradable (coal tar, phenols) and non-biodegradable (heavy metals, recalcitrant cyanide complexes, sulphates) chemicals. Soil containing non-biodegradable pollutants was encapsulated, while a bioremediation programme was developed and implemented for the 30 500m³ soil containing the coal tar and phenols. Before remediation was undertaken, by BioTreatment Ltd. of Cardiff, target reductions in pollutant levels were agreed. The process was developed in three phases: (1) selection of effective pollutant-degrading micro-organisms; (2) optimization of chemical additions to the soil using laboratory microcosms, and (3) optimization of physical field conditions. Micro-organisms able to grow on phenols or polycyclic aromatic hydrocarbons (PAH) were isolated by batch enrichment culture and the bio-availability of these chemicals in the soil was increased by the addition of surfactants. The bioaugmentation of the soil by the isolated organisms was made to accelerate initial stages in pollutant degradation, while the rationale of bioenrichment with nutrients was an enhanced degradation of breakdown products by the indigenous microflora. Regular tillage of the soil proved to be a more effective means of aeration than passive or forced aeration via coils buried in the soil. Full scale treatment of the site started in the summer of 1986 and was complete in just over one year. The contaminated soil was screened and arranged in beds and subsequently inoculated by means of an agricultural spraying machine. Rotavation, watering and booster applications of organisms and supplements occurred during the remediation period. Once target concentrations had been achieved, and subjected to independent validation, the treated soil was replaced in the site and compacted. Mean values for the 16 priority PAHs and phenols before and after bioremediation were validated as 22 000 and 148 mg kg^{-1}, and 205 and less than 3 mg kg^{-1}, respectively. Overall reclamation of the site was estimated to have cost 80–85% of the cost of landfill disposal. A full account of this and similar bioremediation projects is provided by Bewley *et al.* (1989).

Recent experience in Prince William Sound suggests that bioremediation can also be effective in large open environments where ecosystem engineering is severely constrained. The U.S. EPA made an extensive evaluation of bioremediation following the Exxon Valdez oil spill in Alaska (see Atlas and Pramer (1990), for brief report) and showed that bioenrichment along 70 miles of shoreline produced at least a doubling in oil biodegradation. This project showed, predictably, that in situations where control of the environment can only be minimal, seasonal climatic changes have a major impact on remediation rates; and, that the application of a novel technology requires a sensitive touch. On the one hand, laboratory and microcosm studies were important in obtaining support for bioremediation at company, state and federal levels; while on the other, assurance had to be given that bioremediation would not jeopardize the fishing industry.

In the U.S. BioTrol Inc. has developed successful turn-key treatment systems for the bioremediation of soil and water which have been applied to petrol, oil, PAH and pentachlorophenol pollution. The aqueous treatment operation is based on a fixed film bioreactor and its effectiveness is shown convincingly in the remediation of groundwater polluted with pentachlorophenol (PCP). Micro-organisms capable of degrading PCP were isolated from experimental streams within which competent bacteria developed on rock surfaces (Frick *et al.*, 1988). Scale-up column reactors containing biofilms were very effective in degrading PCP and associated, non-target pollutants including PAHs (Pflug and Burton, 1988). In a typical operation contaminated water is pretreated to remove floating oil or suspended solids followed by pH adjustment, nutrient addition and temperature control. Residence times in the bioreactors are of the order of 2h and units are available in various sizes up to 1000 U.S. gallons per minute capacity.

Treatment costs are usually less than 10 cents per gallon and are highly competitive with alternative technologies such as carbon stripping and chemical oxidation. BioTrol's soils treatment process is based on a soil washing system; it is mobile and the large scale system (20 tonnes per hour) can be accommodated on six lorries. The contaminated soil is screened and slurried and following a multistage washing protocol the contaminated water and fine particles streams are treated in bioreactors.

10.6 PRESENT AND FUTURE PRIORITIES

Many of the constraints on biotechnological approaches to hazardous chemical treatment have been publicized repeatedly (Bull, 1980, 1989; Hirschhorn, 1988; Staps, 1989). A short resumé of some constraints and priorities will conclude this chapter.

1. *In situ* bioremediation has great attraction for site clean-up but it can only be applied to biodegradable chemicals in soils with high permeability.
2. Standards for remediation operations are in need of refinement. As Bewley (1990) has pointed out the environmental hazard posed by specific chemicals vary according to the physicochemical nature of the chemical (thence its availability) and the physicochemical nature of the environment; moreover, the quantification of pollutants such as PCBs, coal tar, PAHs and mineral oil in environmental samples often poses difficulties. This problem notwithstanding, comprehensive pollutant audits on waste streams and contaminated sites are essential if effective biotreatment processes are to be developed.
3. The development of sensitive, rapid and robust biosensors (immunoassays, enzyme inhibition, bioadsorption) are needed to complement chemical analyses for hazardous materials.
4. The clean-up of very complex waste sites is a major challenge and will require application of several types of remediation treatment.
5. The usefulness (and acceptability) of genetically manipulated microorganisms for treating hazardous chemicals remains to be proven under realistic operating conditions. At present, the selection of naturally occurring micro-organisms and communities is generally favoured.
6. The scientific base for bioremediation technologies must be strengthened if rationally designed and optimized processes are to be introduced. Substantial gaps exist in the understanding of microbial ecology and physiology, genetic expression and site engineering.
7. There is a persistent need for demonstration processes and their evaluation.
8. Greater emphasis should be placed on integrated pollution control and the particular contributions of biotechnologies in constraining and reducing pollution, and thereby reducing the burden on industry and the taxpayer.

9. Innovative biotreatment technologies must be responsive to public concerns.

REFERENCES

Agg, R. and Zabel, T. (1989) Environmental quality objectives and effluent control. *Chem. Indy.* **14**, 443–447.

Atlas, R.M. and Pramer, D. (1990) Focus on bioremediation. *Am. Soc. Microbiol. News.* **56**, 352–353.

Bewley, R. (1990) Setting standards for restoration of contaminated soil and groundwater. *Chem. Indy* **24**, 354–357.

Bewley, R., Ellis, B., Theile, P., Viney, I. and Rees, J. (1989) Microbial clean-up of contaminated soil. *Chem. Indy.* **23**, 778–783.

Bull, A.T. (1980) Biodegradation: some attitudes and strategies of micro-organisms and microbiologists. In *Contemporary microbial ecology* (ed. D.C. Ellwood *et al.*), pp. 107–136. London: Academic Press.

Bull, A.T. (1989) Mixed culture microbiology and technology. In *Mixed and multiple substrates and feedstocks* (ed. G. Hamer, T. Egli & M. Snozzi), pp. 55–70. European Federation of Biotechnology; Konstanz: Hartung-Gorre-Verlag.

Bull, A.T., Holt, G. and Hardman, D.J. (1988) Environmental pollution policies in light of biotechnological assessment. In *Environmental biotechnology* (ed. G.S. Omenn), pp. 351–371. New York: Plenum Publishing Corporation.

De Brauw, D.J. (1989) The Dutch approach. In *Environmental regulation in the European Community. Cost, planning & strategic issues for industry*, pp. 1–18. London: Legal Studies & Services Ltd.

Frick, K.T.D., Crawford, R.L., Martinson, M., Chesand, T. and Bateson, G. (1988) Microbiological clean-up of groundwater contaminated by pentachlorophenol. In *Environmental biotechnology* (ed. G.S. Omenn), pp. 173–191. New York: Plenum Publishing Corporation.

Gieseler, G. (1987) Contaminated in the EC. Report issued by Ministry of Research & Technology, pp. 1–216. Bonn.

Her Majesty's Inspectorate of Pollution (1989) The management of hazardous waste in the United Kingdom. London: H.M.S.O.

Hill, R.D. (1988) *Status of the Superfund Innovation Technology Evaluation (SITE) Program*, pp. 516–520. Superfund 88, Proceedings of the 9th National Conference. Silver Spring, MD: The Hazardous Materials Control Research Institute.

Hirschhorn, J.S. (1988) Superfund strategies and technologies: a role for biotechnology. In *Environmental biotechnology* (ed. G.S. Omenn), pp. 419–427. New York: Plenum Publishing Corporation.

Jones, K.C. and Peace, E.A. (1989) The Ames mutagenicity assay applied to a range of soils. *Chemosphere* **18**, 1657–1664.

Nicholas, R.B. (1987) Biotechnology in hazardous waste disposal: an unfulfilled promise. *Am. Soc. Microbiol. News* **53**, 138–142.

Nunno, T., Hyman, J., Spawn, P., Healy, J., Spears, C. and Brown, M. (1989) *Assessment of international technologies for Superfund applications – technology identification and selection*. U.S. EPA/600/2–89/017, pp. 1–283. Washington, D.C.: U.S. Government Printing Office.

Pflug, A.D. and Burton, M.B. (1988) Remediation of multimedia contamination from the wood-preserving industry. In *Environmental biotechnology* (ed. G.S. Omenn), pp. 192–201. New York: Plenum Publishing Corporation.

Slater, J.H. and Bull, A.T. (1982) Environmental microbiology: biodegradation. *Phil. Trans. R. Soc. Lond.* B **297**, 575–597.

Staps, S. (1989) Biorestoration of contaminated soil and groundwater. *Chem. Indy* **23**, 581–584.

Superfund 88 (1988) *Proceedings of the 9th National Conference*, Silver Spring, M.D.: The Hazardous Materials Control Research Institute.

U.S. EPA (1988) *Superfund innovative technology evaluation. Program description.* Washington, D.C.: U.S. EPA.

World Commission on Environment and Development (1987) *Our Common Future.* Oxford University Press.

11

Degradation of organic wastes

G.K. Anderson and M.B. Pescod

11.1 INTRODUCTION

This paper will be concerned with treatment technologies capable of reducing organic pollutant concentrations by means of their molecular dissimilation. Issues related to environmental protection now regularly make headline news, and the 'environment' is now high on political agenda throughout the world. International issues, such as acid rain, global warming and the hole in the ozone layer have focused public attention to the critical need for mankind to live in harmony with his environment by the proper disposal of wastes. Though the initial momentum behind this growing environmental awareness was created by publicity given to the major global issues, the importance of the message has lead to an increasing demand for closer scrutiny of all activities which might impinge on the environment. One activity of particular concern is that of the disposal of organic wastes in the environment, since such materials may be responsible for a reduction in natural oxygen levels, be toxic, or simply be aesthetically displeasing. To simplify the overall picture of degradation of organic wastes, they will be considered in their separate physical forms, mainly liquid, solid and gaseous.

11.2 LIQUID ORGANIC WASTES

11.2.1 Sources

Liquid wastes containing organic matter may be divided broadly into domestic sewage and industrial wastewater. A river pollution survey of England

The Treatment and Handling of Wastes
Edited by A.D. Bradshaw, Sir Richard Southwood and Sir Frederick Warner
Published in 1992 by Chapman & Hall, London, for The Royal Society
UK ISBN 0 412 39390 5, USA ISBN 0 442 31461 2

and Wales (HMSO, 1978) in 1975 showed that, of industrial discharges containing Biochemical Oxygen Demand (BOD), the main contributors on a volumetric basis were the paper (30%), chemical (25%) and food and drinks (20%) industries. Over 80% of sewage (domestic and industrial) produced in England and Wales is subject to secondary (biological treatment), and in 1980 it was estimated that 94% of the polluting load (BOD) of sewage was removed by treatment before being discharged into inland rivers. For discharges into tidal waters, the proportion removed is over 50% (HMSO, 1973).

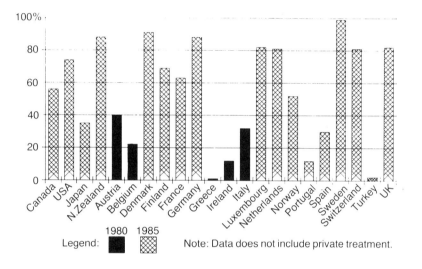

Figure 11.1 Percentage of the population served by wastewater treatment plants.

On a global basis, as shown in the statistics compiled by OECD (1987) and summarized in Fig. 11.1, the picture is very variable. For example, the rapid expansion of population in Japanese cities during the 1960s and 1970s has lead to the country currently having insufficient treatment facilities to accommodate the production of wastes. On average, only 34% of the country's wastewaters are treated before release (Preston, 1989). A breakdown of pollution loadings flowing into major rivers in Japan showed that 55% of the polluting load is domestic, whilst 35% originates from industry, though this proportion has been falling since the enactment of laws in 1973 to limit industrial discharges (Preston, 1989).

Though the various wastewaters can differ markedly in the nature of their organic constituents, a common set of processes tends to be used for their degradation. This discussion will use the food processing industry as a model to show current technology and possible developments in the treatment of liquid wastes in the third millenium.

Table 11.1 Volume and polluting load of wastewaters from the food industry

Industry	BOD	SS	Grease	Water
Meat processing[a]				
Killing	5.8	4.7	2.5	5–20
Processing	5.7	2.7	2.1	10–40
Frozen foods[b]				
Peas	8.9–20.6	2.7–13.0	2.5	8.0–19.0
Beans	1.3–5.4	1.3–2.2	—	20.9–40.0
Poultry[c] (average)	13.6	6.9	0.6	14.8
Dairy[b] (average)	5.9		—	2.4
Wheat processing[b]	90	5	—	20
Potato processing[b]	264	—	—	14–23
Ice cream[b]	1.4–8.3		0.4–1.25	
Margarine[b]	1.2–4.0	—	0.6–2.0	1
Edible oil refining[b]	2.5–6.8	—	1.25–3.0	2–50

[a] kg t^{-1} of live weight or m^3 t^{-1} of live weight.
[b] kg or m^3 t^{-1} of product.
[c] kg or m^3 per 1000 birds.

11.2.2 Characteristics of food-processing wastewaters

The characteristics of food-processing wastewaters vary considerably from one type of operation to another. In general, however, the main constituents are animal, fish or vegetable fat, protein and carbohydrates. Typical characteristics from some common food operations are given in Table 11.1.

11.2.3 Discharge controls on food-processing wastewaters

As food processing wastewaters contain essentially biodegradable, non-toxic material the relevant standards that are applicable to such discharges are those normally applied to the discharge of treated domestic sewage to water courses, i.e.:

biochemical oxygen demand (BOD)	20 mg l^{-1}
suspended solids	30 mg l^{-1}
ammoniacal nitrogen	10–20 mg l^{-1}

The above standards can vary and are determined by the condition and use of the receiving water and the dilution that is available.

In the case of discharges to sewer, then the fat content of the wastewater is often controlled. Fatty matter in wastewater can be broadly classified by the following self-explanatory definitions: total fatty matter (TFM), separable fatty matter (SFM) and non-separable fatty matter (NSFM). The fat limits imposed on industrial discharges to sewers varies considerably from one part of the country to another but normally total fatty matter is used as the controlling parameter. Typical examples are given in Table 11.2.

Table 11.2 Some total fatty matter standards applied to specific food industry discharges to sewerage systems

Type of operation	Total fatty matter (TFM) (mg/1)
Frozen foods (A)	100
(B)	No limit
(C)	50
Edible oil refinery (D)	150
(E)	500
Meat processing	200
Ready meals	400
Ice cream	400

11.3 MECHANISMS FOR CONTROL

11.3.1 Pretreatment

In general, biological processes are used for the degradation of organic matter in domestic and industrial wastewaters. However, such processes cannot be considered in isolation since, in most cases, it is necessary to pre-treat the wastewater to protect the biological system. Typical pre-treatment processes include the following:

1. screening. Removes discrete particulate matter. Bar screens and wedge wire or perforated rotating drums remove solids down to 1mm;
2. separable fat removal. Whatever additional treatment is required, it is often desirable first to remove separable fatty matter. Generally, gravity fat-separators, etc.;
3. flow balancing. To improve the efficiency of biological treatment system, flow and load balancing may be necessary, consisting of a holding tank, the contents of which are mixed. The balancing tank is an integral part of the treatment process for many wastewaters and has a significant effect on overall treatment efficiency;

4. sedimentation. Suspended particles may be removed by sedimentation (or by filtration in some cases) although it is debatable whether settlement is justifiable in many cases undesirable fermentation may take place during the retention period;
5. chemical treatment. A wide range of chemical pre-treatment systems are available to which careful consideration should be given before to biological degradation These include neutralization to pH 6–8, coagulation and flocculation.

In general, chemical treatment is a partial treatment process removing insoluble components of the wastewater, such as emulsified fat. The process usually removes little of the soluble BOD, hence the overall BOD removal efficiency depends greatly upon the ratio of non-soluble to soluble components in the wastewater. Because of the nature of the processes involved, successful treatment depends upon close supervision, and the minimum of variation in raw wastewater characteristics. Labour costs can, therefore, be an important element of the operating costs of the chemical treatment process.

Financially, a chemical treatment process may be characterized as one of moderate capital cost but very high operating costs, due primarily to chemical consumption, manpower and sludge disposal.

11.3.2 Aerobic biological treatment

Most effluents arising from food industries are biodegradable and totally amenable to treatment by well designed biological processes. Activated sludge and biological filter processes, originally developed for the treatment of domestic sewage, have been applied to food industry effluents, often with complete success.

When applying biological processes to the treatment of industrial effluents however, care must be taken to completely characterize the effluent to be treated and perform treatability studies to determine design parameters for the full-scale plant. The basic reaction within an aerobic treatment plant is as follows:

$$\text{Organic material} + O_2 \xrightarrow{\substack{\text{(cells)} \\ \\ \text{(nutrients)}}} CO_2 + H_2O + \text{new cells}$$

At the same time bacteria undergo progressive autoxidation of their cell mass as described by:

$$\text{cells} + O_2 \rightarrow CO_2 + H_2O + NH_3$$

The various types of aerobic biological treatment systems that have been used in the food industry are listed in table 11.3. The list has been divided

Table 11.3 Aerobic treatment systems

Suspended culture systems	Attached culture systems
Lagoons	Low rate biological filtration
Activated sludge	High rate filters

into those systems in which the treatment micro-organisms are in free suspension and those in which the organisms are attached to an inert support medium over which the effluent flow passes during treatment.

Lagoons and low rate filters have limited industrial application within the U.K. because of the large areas of land which are normally required for such treatment processes.

11.3.3 Activated sludge

The activated sludge system consists of an aqueous suspension of micro-organisms in contact with wastewater and oxygen, removing organic substrates from the waste by absorption, oxidation and synthesis of raw cell material. In its simplest form it can be regarded as a reaction vessel and settlement tank in series with a suitable method of sludge return from the settlement tank to the aeration tank inlet.

The contents of the reaction vessel are referred to as mixed liquor suspended solids (MLSS) or mixed liquor volatile suspended solids (MLVSS), and consist mostly of micro-organisms and inert and non-biodegradable suspended matter. In general, food processing wastewaters are deficient in nitrogen, therefore the bacteria which predominate in the reactor are those responsible for carbonaceous BOD removal. The design of the process is consequently based on providing a controlled environment for these micro-organisms.

Briefly, the general design considerations are as follows.

1. Nutrient requirements. The ratio of BOD:nitrogen:phosphorus must be not more than 100:5:1 to ensure sufficient nutrient supply. Often, for food processing wastewaters, this requires continuous dosing facilities for soluble inorganic salts containing nitrogen and phosphorus.
2. pH. The optimum pH range is 7.0–8.0, thus pH adjustment of the wastewater may be required prior to the reactor. Satisfactory performance cannot normally be obtained outside the range 6.0–9.0.
3. Temperature. Activated sludge is usually operated at ambient temperature. Systems using pure oxygen as the oxygen source may operate at higher temperatures, in the range of 30–40 °C.
4. Fatty matter. High levels of fatty matter absorbing on the microbial flocs, which constitute the MLSS, can adversely affect the activated

sludge process by reducing the rate of oxygen transfer to the micro-organisms. Generally, the maximum allowable loading of fats, oils and greases (FOG) is in the range of 0.1–0.25 kg FOG kg^{-1} MLSS/day. As an approximate guide, if the raw wastewater FOG (or TFM) concentration is greater than 50% of the total BOD concentration, then chemical treatment should be considered as a pretreatment to remove the fatty matter before the activated sludge process.

5. Loading rate on the reactor. There are a number of approaches to reactor design but they all generally result in a reactor loading of 0.1–0.3 kg BOD kg^{-1} MLVSS/day calculated from the average daily BOD load. In the case of carbohydrate rich wastewaters, the lower loading is often applied to avoid the undesirable condition of 'bulking' in which the micro-organisms cease to flocculate and settle in the sedimentation tank. At the very least, this results in a significant deterioration in effluent quality, and at the worst it can cause complete loss of the essential micro-organisms. For systems operated with a MLSS concentration in the range 2000–4000 mg l^{-1}, the biological loading rate quoted above equates to an organic loading rate of 0.2–1.2 kg BOD m^{-3} reactor/day.

6. Configuration of reactor. Currently in order to prevent the formation of a 'bulked' sludge and plug flow, baffled reactors are favoured to completely mixed reactors.

7. Aeration and mixing. Aeration and mixing are generally carried out as a combined function with the energy input being determined by the aeration requirements. There are many different methods of aeration, including turbine type surface aerators, brush type aerators, coarse bubblers, fine bubblers, and pure oxygen injection. The aerators must not block, but they are selected primarily on efficiency in terms of mass of dissolved oxygen achieved per kilowatt hour (kwh) consumed. On average, an aerator efficiency of 2 kg O$_2$ kwh^{-1} might be expected and BOD removal related to power consumption would be approximately 0.5–0.6 kg BOD kwh^{-1}.

8. Excess sludge production. Excess biological sludge is produced with a yield ranging from 0.3 to 0.5 kg SS/kg BOD applied. Treatment and disposal of this by-product sludge can add up to 50% of the overall treatment cost and is therefore planned as an integral part of any wastewater treatment scheme. The sludge is generally amenable for disposal on to farmland, therefore in rural locations, sludge treatment might consist simply of storage and gravity thickening. In the urban environment, the cost of haulage of liquid sludge to farmland can become prohibitive, and necessitates further sludge treatment in the form of dewatering by centrifugation or filtration.

9. Final clarifier. Separation of the micro-organisms from the treated effluent is generally carried out in a circular, radial flow vessel with a shallow conical base and a continuous scraper. The vessel is usually

2.5–3.5m deep and the surface area is calculated from the peak flow by using an overflow rate of approximately 12 m^3 m^{-2} d^{-1}.
10. Degree of treatment. The activated sludge process is used to provide a high degree of treatment with in excess of 95% BOD removed. For dilute wastewaters, i.e. less than 1000 mg/1 BOD, the process is capable of providing an effluent with BOD and SS concentrations of less than 20 and 30 mg l^{-1}, respectively, which would be suitable for discharge to river.

11.3.4 High-rate filters

In the biological filter the micro-organisms are attached as a thin film on to the surface of an inert supporting medium (generally plastic) which is packed into a tower or a tank. This 'reactor' is fitted with a fixed channel or rotating arm distributor at the top to permit even distribution of wastewater, and air vents at the bottom to allow easy passage of air up through the media bed. Wastewater applied to the top layer of the bed percolates down the voids between the media, and air flows upwards because of the natural thermal gradient set up by the incoming wastewater.

In the presence of organic material and air, a microbial slime develops on the media, and the bacteria within this slime layer removes the organic material from the wastewater. As the thickness of the slime layer increases due to microbial growth, it falls from the media (sloughs) and is carried down through the bed and out with the treated wastewater. This biological sludge is then removed from the treated wastewater in a gravity sedimentation tank. The sludge, known as humus sludge, is treated and disposed of in the same way as that described for activated sludge.

Pretreatment. The pretreatment requirements for high rate filters are similar to those for activated sludge in terms of pH, temperature, fatty matter and nutrients.

Loading rate and treatment efficiency. High rate filters are employed where a high quality effluent is not required i.e. as pretreatment systems or for treatment prior to being discharged to sewer. Loading rates range from 3–10 kg BOD m^{-3} d^{-1} and result in treatment efficiencies within the range of 70–30% BOD removal. Apart from specifying the desired organic loading rate, it is also necessary to ensure a minimum wetting rate to provide the attached micro-organisms with a continuous supply of nutrients, moisture and air. Typical wetting rates range from 0.5 to 1.5 m^3 m^{-2} media surface per hour depending upon the media type.

Recycle. For many wastewaters, the wastewater flow must be supplemented by recycle of treated effluent in order to achieve the desired wetting rate. However, unlike the activated sludge process, there is no need for recycle of the micro-organisms, therefore in this case the recycle may be taken from points either upstream or downstream of the final clarifier.

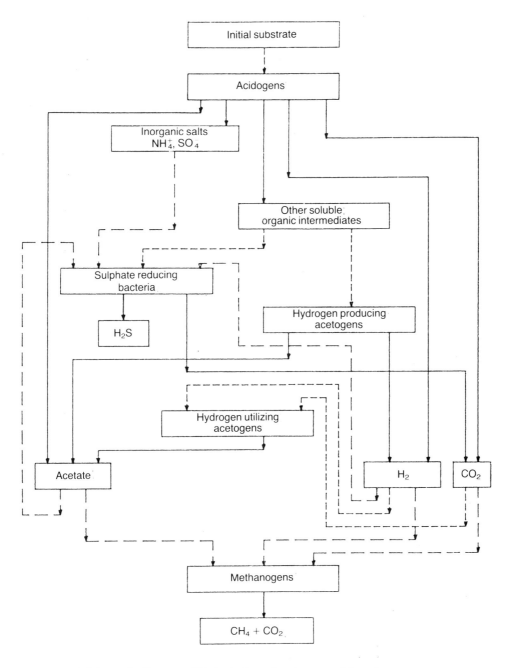

Figure 11.2 Schematic diagram of biological reactors in anaerobic systems.

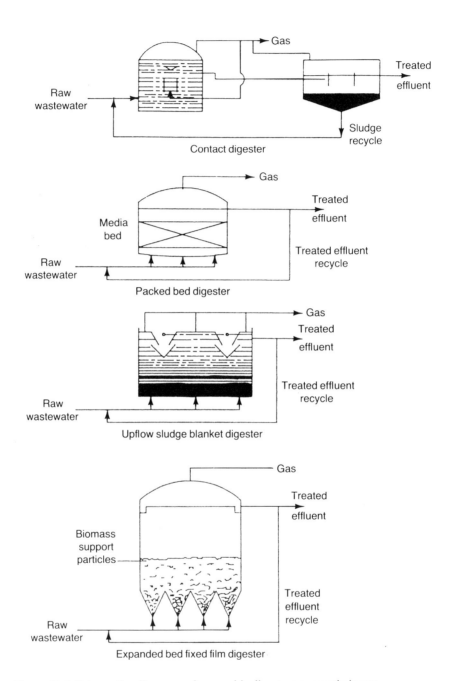

Figure 11.3 Schematics diagrams of anaerobic digesters currently in use.

Recycle from upstream of the clarifier has the advantage of reducing the required surface area of the clarifier, and reducing odour nuisance from the reactor, though care must be taken to ensure that sloughed biomass, when recycled, does not block the distributors and surface of the media bed.

Excess sludge production. The yield of excess sludge from high rate filtration ranges from 0.5 to 1.0 kg SS kg^{-1} BOD. This excess sludge is generally difficult to dewater, and its treatment and disposal can add significantly to the overall cost of wastewater treatment.

11.3.5 Anaerobic biological treatment

Anaerobic biological digestion as applied to the treatment of wastewaters has many significant advantages over the conventional aerobic processes and is a technique which is becoming increasingly used. Anaerobic digestion is a microbial fermentation by which, in the absence of high levels of sulphate or nitrate, organic matter is converted into carbon dioxide and methane as shown schematically in Fig. 11.2. Anaerobic digestion has long been practised as a stabilization process for waste sewage treatment sludges but was initially seriously considered only for industrial application in the treatment of very high strength wastes.

During recent years however, energy and sludge disposal costs have served to highlight the advantages of anaerobic systems especially for industrial applications. This has resulted in a considerable effort by academics and industry into achieving a better understanding of the process. Currently a number of processes are commercially available, e.g. contact process, upflow sludge blanket, packed bed reactor, and the fluidized bed process, shown schematically in Fig. 11.3 (as 'Expanded bed').

The main advantages of anaerobic digestion are:

1. very low sludge production;
2. a substantial proportion of the organic compounds is converted to a gas of high calorific value (methane);
3. low consumption of energy (cf. aerobic processes);
4. process operates at high organic loading rates;
5. no environmental nuisance as the process is totally enclosed;
6. ability of micro-organisms to be dormant for several months and then be fully operational within one week of start-up (of great value when seasonally produced wastewaters are to be treated).

Though detailed design requirements differ for each reactor configuration, general process design considerations are briefly as follows.

1. Nutrient requirements. The ratio of BOD:nitrogen:phosphorus must not be more than approximately 100:2:0.5 to ensure sufficient nutrient supply. This low nutrient requirement can very often be satisfied by the

raw wastewater from food processing, though careful analyses should always be made to ensure that this is the case.

2. pH. Satisfactory operation may be obtained within the range of 6.5 to 8.5, but the optimum range is 6.8 to 7.5. Adjustment of the raw wastewater pH may therefore be necessary. However, unlike aerobic treatment systems, the formation of short chain fatty acids during storage and balancing can be tolerated by anerobic systems in certain cases without the need for pH adjustment.

3. Temperature. Most anaerobic processes currently operate in the mesophilic range of 30–37°C. At lower temperatures, the reaction falls off sharply, resulting in the need for larger reactors. Energy for heating is best obtained from low grade sources within the factory, which then leaves the by-product methane gas to be used as a potential high grade energy source for the production of hot water, steam or electricity.

4. Fatty matter. A potentially important characteristic of the anaerobic process is its ability to assimilate relatively high levels of fatty matter. This would be a new area of application for the process, which can offer significant economic benefits by replacing chemical treatment as a pre-treatment process for fat-bearing wastewaters.

5. Reactor loading rate and treatment efficiency. Depending upon the reactor configuration, loading rates can range from approximately 1.5–8.0 kg BOD $m^{-3} d^{-1}$, with corresponding BOD removal efficiencies ranging from greater than 90% to 70%.

6. Gas production and energy requirements. Energy may be required for heating, pumping and mixing. Although heat may be recovered from the treated effluent, in many cases the factory has a low grade source of heat (often the wastewater itself) which may be used for process heating. In these cases, the energy requirement is very low, being simply pumping and mixing. Of course, anerobic systems usually produce a methane rich gas as a by-product. The gas normally contains between 60–80% methane with 40–20% carbon dioxide and traces of hydrogen sulphide. The gas production rate is approximately 0.9–1.0 $m^3 kg^{-1}$ BOD applied and has an approximate calorific value of 25 MJ m^{-3}. Although of potential value, this high energy by-product can rarely be used as sole justification of anaerobic digestion, the process being primarily an effluent treatment system.

11.4 TECHNOLOGY DEVELOPMENTS IN LIQUID WASTE TREATMENT

11.4.1 General

It is plain from the OECD data summarized in Fig. 11.1, that a significant fraction of the wastewaters in the world's leading industrial countries do not yet receive secondary treatment. The direction of legislative developments is now irreversibly towards a dramatic increase in secondary treatment. Where land space permits, it would seem to be inevitable that well proven aerobic biological treatment systems will be adopted. Developments to fine-tune these systems are likely to be aimed at greater automation and on-line monitoring. Continued improvements in suspended solids and dissolved oxygen monitoring and in-line organic carbon analysis should lead to the possibility of remote control of the reactor performance through control of the food to micro-organism ratio or mean solids retention time. Such information can also be used to minimize power consumption by closely tailoring aeration power to oxygen demand.

Whilst current technology will continue to be improved by such details, the key areas for development of biological treatment systems must include improved biomass separation; anaerobic pre-treatment to reduce by-product sludge; microbial enhancement.

11.4.2 Biomass separation

A central factor in the successful operation of biological wastewater treatment systems, is the ability to control the loss of biomass. The need for more extensive secondary treatment in locations such as Japan will give rise to improvements in biomass separation. The main driving force in these cases is the lack of available land. Gravity separation of biomass in conventional activated sludge systems limits the biomass concentration in the reactor to a maximum of approximately 4000 mg l^{-1}. Doubling the biomass concentration, perhaps through membrane separation techniques, would halve the reactor volume.

Membrane separation of biomass could lead to improvements in system reliability and the possible use of higher food to micro-organisms (F:M) ratios. Hitherto this could not be applied on many industrial wastewaters the dreaded filamentous bulking problem that results in catastrophic biomass loss from gravity solid–liquid separators. A doubling in F:M will lead to a further halving of the reactor volume. Thus successful development of micro-filtration and ultrafiltration membranes will revolutionize the activated sludge process by vastly reducing the reactor volume.

11.4.3 Anaerobic pre-treatment

Anaerobic biological systems are already being used extensively in the sugar industry. The increasing restrictions being placed on the disposal of solid wastes, including excess biological sludge from aerobic systems, will lead to more widespread use of anaerobic processes, which produce between 15 and 20% of the by-product sludge of the aerobic alternative. The design of anaerobic systems is more of a chemical engineering than civil engineering approach which dominates in municipal wastewater treatment, therefore some re-orientation in approach will be required. However, anerobic systems, having suffered early development problems in the 1970s are now receiving widespread attention in industry. In hot climates, where the sewage is warm, anaerobic systems could well extend in the municipal field as a pretreatment process for mixed domestic and industrial wastewater, thus partly alleviating sludge treatment and disposal problems.

In terms of the design of anaerobic systems themselves, the immobilization of biomass in an expanded or fluidized bed has great potential. The extremely high concentrations of biomass which can be retained within such reactors (>100000 mg l^{-1}SS) offers the potential to reduce the anaerobic reactor volume by a factor of 2 over other high-rate systems. More research is required, however, to improve the understanding of nozzle design and media effects on attachment and bed stability.

11.4.4 Microbial enhancement

The use of specific bacteria for the treatment of wastewaters continues to be the process designer's dream. Instead of, as at present, being at the mercy of a heterogeneous biomass enriched from a seed sludge obtained from the geographically nearest reactor, the process designer would wish to specify and control the micro-organisms within the system. This occurs at present, to a certain degree, with the use of freeze-dried bacteria, to start-up aerobic systems and to enhance fat degradation under certain conditions, and the use of granular structures of anaerobic bacteria for the start-up of the upflow anaerobic sludge blanket process. However, the introduction of environmental charges for the discharge of semi-recalcitrant organic chemicals is likely to result in an upsurge of research into the use of specific microorganisms as a pre-treatment stage of conventional systems. Apart from the isolation, genetic manipulation, culturing and 'biomass-banking' of such organisms, research will need to be directed towards a means of immobolizing or retaining the desired bacteria within a suitably engineered system. Whilst membrane separation techniques should be of use for this purpose, immobilization systems, such as encapsulation may also have a role to play.

11.5 SOLID ORGANIC WASTES

The generation of solid wastes is the inevitable result of man's activities. There is, however, no standard solid waste; in fact no two wastes will be identical. It is also often difficult to compare wastes from different areas since sampling and analytical methods used to obtain the original data are not always known. In recent years, much greater attention has been paid to the planning of solid waste management on a national and regional scale. With the general increase in volume of waste in urban areas and the changing composition over time, it is essential to plan for future trends and waste arisings. In the present climate of environmental concern it will be prudent of all solid waste regulation authorities to consider the likely demands to be placed upon them by the pubic and the resources will be necessary to meet those demands.

11.5.1 Effects of solid waste

Solid waste management is an essential urban operation which is provided to achieve protection of public health; promotion of hygiene; recycling of materials; avoidance of waste; reduction of waste quantities, and reduction of emissions and residuals. Solid waste is likely to contain human pathogens from a wide range of sources and its handling and disposal can present the same potential for disease transmission as the collection and disposal of excreta and sewage. Toxic and hazardous substances which are generated by both the public and industrial sector may also be present. In addition to health and safety problems, solid waste collection and disposal systems will have broader environmental impacts which require attention. Any solid waste treatment or disposal system will have a visual impact on the surroundings and consideration must always be given to minimizing visual intrusion.

11.5.2 Types of waste

Solid waste may be broadly classified into: domestic; commercial; industrial.

(a) Domestic Waste

The composition of domestic waste must be known to evaluate the required treatment and disposal systems. The proportion of organic material present may determine if composting, incineration or digestion are appropriate methods. In addition, evaluation of the feasibility of incineration depends upon the chemical composition of the waste. Table 4 shows the changing composition of solid waste in the U.K. from the 1930s to 1980s.

(b) Commercial solid wastes

The composition of commercial solid wastes depends entirely upon the source. They may include: office buildings; restaurants; markets, and hotel and motels.

(c) Industrial solid waste

Since the term industrial solid wastes could cover an extremely wide spectrum of waste materials, it is valuable to classify wastes into three broad

Table 11.4 Compositions of domestic solid waste – U.K. national average

Classification		1935	1963	1967	1968	1969	1970	1972	1973	1974	1975	1976	1977	1978	1979	1980
Screenings (20 mm) Dust/Ash	(%)	56.9	38.5	31.0	21.9	17.2	14.9	19.9	18.7	19.8	18	18	14	11	12	14
	(kg)	9.7	5.5	4.0	2.9	2.2	2.0	2.3	2.2	2.1	7.1	1.8	1.4	1.2	1.4	1.6
Vegetable and putrescible	(%)	13.7	14.1	15.5	17.6	19.5	24.5	19.5	18.1	21.3	20	19	25	29	24	25
	(kg)	2.3	2.0	2.0	2.3	2.5	3.3	2.3	2.1	2.3	2.4	2.0	2.5	3.2	2.6	2.8
Paper & board	(%)	14.3	23.0	29.4	36.9	37.9	36.8	30.5	32.7	26.8	30	24	26	27	29	29
	(kg)	2.5	3.2	3.8	4.9	4.9	5.0	3.6	3.8	2.8	3.4	2.4	2.7	3.0	3.2	3.2
Metals	(%)	4.0	8.0	8.0	8.9	9.7	9.2	8.7	8.8	8.5	8	8	9	7	8	8
	(kg)	0.7	1.1	1.0	1.2	1.2	1.3	1.0	1.0	0.9	9.9	0.8	0.9	0.8	0.9	0.9
Textiles & manmade fibre	(%)	1.9	2.6	2.1	2.4	2.3	2.6	3.0	3.1	3.5	3	4	3	4	4	3
	(kg)	0.3	0.4	0.3	0.3	0.3	0.3	0.4	0.4	0.4	0.3	0.4	0.3	0.4	0.5	0.3
Glass	(%)	3.4	8.6	8.1	9.1	10.5	9.0	10.4	10.5	9.5	9	9	11	9	10	10
	(kg)	0.5	1.2	1.1	1.2	1.3	1.2	1.2	1.2	1.0	1.1	0.9	1.1	0.9	1.1	1.1
Plastic	(%)			1.2	1.1	1.4	1.4	1.9	2.0	2.9	4	5	5	5	7	7
	(kg)			0.2	0.2	0.2	0.2	0.2	0.2	0.2	0.3	0.5	0.5	0.6	0.8	0.8
Unclassified	(%)	5.8	4.9	4.7	2.1	1.5	1.6	6.1	6.1	6.9	8	14	7	6	6	4
	(kg)	1.0	0.7	0.6	0.3	0.2	0.2	0.7	0.7	0.7	0.9	1.4	0.7	0.6	0.6	0.5
Total per house per week	(kg)	17.0	14.1	13.0	13.3	12.8	13.5	11.7	11.6	10.7	11.6	10.2	10.1	10.9	11.1	11.2
Density	$(kg\ m^{-3})$	290	200	160	157	143	146	153	151	161	164	152	126	141	141	147

categories: non-hazardous industrial solid waste; hazardous waste; hospital waste.

Non-hazardous industrial solid waste

Many industries produce solid waste materials from fabrication, chemical, refining, quarrying, power generation and other processes. If they are classified as non-hazardous then they may be stored, collected, treated and disposed of along with urban solid waste by either private or public sector operations. The type and quantity will be dependent upon the nature of the industry and the types of production process.

Hazardous wastes

The term hazardous waste is used here to describe a range of materials which often are described by using other terms such as 'difficult waste', 'toxic waste' or 'special waste'. Table 11.5 is a summary of the list of wastes that are

Table 11.5 Classifications of hazardous wastes

Type of waste	Typical example(s)
Inorganic acids	sulphuric acid, nitric acid
Organic acids and related compounds	formic acid, benzoyl chloride
Alkalis	sodium hydroxide, ammonia
Toxic metal compounds	cadmium, mercury, arsenic
Non-toxic metal compounds	titanium
Metals	sodium
Metal oxides	cadmium oxide, beryllium oxide
Inorganic compounds	cyanides, peroxides, chromates
Other inorganic materials	asbestos, slag, silt
Organic compounds	hydrocarbons, phenols, halogenated cleaning compounds, organo metalics
Polymer materials	epoxy resis, latex, ion-exchange resins
Fuels, oils, greases	mineral oil, fuel oil, fats, waxes
Fine chemicals and biocides	pharmaceuticals, biocides
Miscellaneous wastes	organics and inorganics identified by trade name only
Filter materials, treatment sludge, contaminated rubbish	kieselguhr, empty containers, industrial waste sludges
Interceptor wastes, tars, paints, dyes and pigments	printing ink, dyestuff, paint
Miscellaneous waste	tannery waste, cellulose waste
Mineral and food waste	carcasses, glue waste

regarded as 'difficult' and are classified according to the system established by the U.K. Waste Management Paper 'Special wastes' (Department of Environment, 1981). However, the definition of 'Special wastes' is based upon available toxicity data (usually on rats) and this may be unsatisfactory.

Hospital wastes

For many years the World Health Organization has advocated that hospital wastes should be regarded as hazardous wastes. It is now commonly acknowledged that certain categories of hospital (or clinical) waste are among the most dangerous of all wastes arising in the community. As the volume and complexity of health care wastes increase, the risk of transmitting disease through unsatisfactory disposal practices also increases.

11.5.3 Methods for control

Treatment methods for solid waste are used to reduce the mass or volume and to achieve one or more of the following objectives:

1. improvement of its acceptibility in environmental terms;
2. separation and recovery of recyclable material or energy;
3. reduction in transport costs;
4. reduction in volume of required landfill;
5. minimization of overall cost.

Since market forces continually affect the costs of waste treatment and disposal and because the public is becoming more and more environmentally sensitive, it is essential that all treatment and disposal options are kept under review at all times. In terms of degradation of organic wastes, the following options are available: composting; landfill; anaerobic digestion; incineration.

(a) Composting

Composting is an aerobic biological decomposition process which ultimately degrades susceptible organic material to carbon dioxide, water and a stabilized residue, principally humic substances called 'compost'. Bio-oxidation processes are exothermic and substantial quantities of heat are produced in the initial part of the process, causing the temperature to rise. This, in turn, vaporizes the moisture thereby reducing the weight and volume of the substrate by some 50% during the maturation process.

The composting process is carried out by naturally occurring microorganisms which will spontaneously grow in any natural organic waste if the desirable moisture content and aerobic conditions exist.

The high metabolic activity and exothermic processes cause the temperature of the composting mass to rise to above 60°C, thus having a strong selective effect in favour of the thermophilic sporigenous bacteria and inhibit the growth of a wide range of other micro-organisms.

Since composting is an exclusively biological degradation process, all those factors which influence microbial metabolism affect the process, namely: temperature (35–60°C); moisture content (40–50%); oxygen (15–18%); initial C:N 25:1–35:1 ph 5–8.5.

(b) Landfill

Land disposal in the form of sanitary landfill has proved to be the most economical and acceptable method for the disposal of solid wastes. The term 'sanitary landfill' implies an operation in which the wastes to be disposed of are compacted and covered with a large inert material at the end of each day's operation. When the disposal site has reached its ultimate capacity a final layer of cover material is applied. Degradation (decomposition) of the organic matter which takes place is largely anaerobic.

One consequence of the biodegradation is the generation of methane which, even though it is formed at a slow rate must be vented to atmosphere or burnt under controlled conditions. Attempts are being made in many cities to capture this methane for energy.

A further consideration is that of waste entering the landfill which will create leachate which in turn may pollute surface or groundwater if not properly controlled (see Rae and Campbell, Chapter 8).

(c) Anaerobic digestion of solid waste

It is now accepted that a large amount of energy may be removed from urban solid waste in the form of landfill gas. However, a landfill is an inherently inefficient bio-reactor and the potential energy is recoverable only over a long period, estimated to be between 5 and 10 years. The amount of energy that is recovered may be increased by adopting thermal or thermochemical techniques, including incineration, pyrolysis, gasification or the production of refuse derived fuel (RDF).

It is clear that energy production from solid waste may be optimized by including anaerobic digestion into integrated waste separation and materials recovery systems.

An anaerobic digestion system includes the following unit processes:

1. feed preparation, where the classified/sorted waste is slurried with recycle liquor from the dewatering of digested residue, sewage sludge (optional) and any chemicals required to maintain the digester pH and provide the nutrients required for microbial growth;
2. a reactor, in which fermentation takes place either at mesophilic (30–35°C) or thermophilic (50–55°C) temperatures.

(d) Incineration

Incineration is the term used for the combustion of municipal solid waste. In a properly designed and operated incinerator there is a substantial reduction in the volume of waste material to be eventually disposed of. When incinerated, solid waste becomes a sterile ash with a minimum carbon or fat content and may thus be safely disposed of in any appropriate location. The use of solid waste as a fuel, thereby obtaining a measure of recycling as well as disposal, is widely practised in Europe. The organic content, or calorific value of the waste is the determining factor which is strongly influenced by its cellulose and lignin (paper and cardboard) content. The gross calorific value of cellulose is $17500 kJ\ kg^{-1}$ compared to the average gross calorific value of mixed urban solid waste of about $10500 kJ\ kg^{-1}$.

The two most important aspects of solid waste as a fuel are that it has a low calorific value, typically 30–40% of that of an industrial bitminous coal, and a density, as fired, of about $200\ kg\ m^{-3}$.

11.6 GASEOUS WASTES

11.6.1 Sources

To control air pollution it is necessary to understand what the sources of air pollution are, and how they operate. It is possible, in theory, to control air pollution by eliminating the sources. This, however, would have a most disruptive effect on the way in which we live.

Some sources of air pollution are very large and concentrated: they are the large factories, chemical plants, oil refineries and power stations. These, however, contribute only about one third of the total mass of the air pollution burden (except for CO_2). Transporation contributes about 45% with space heating also making a significant contribution. Incineration of solid waste is estimated to add a further 5%.

Within cities it is the multiplicity of small sources, particularly private motor cars that are the main cause of the degradation of air quality.

(a) Effects of gaseous pollutants

After formation, air pollutants are emitted to the atmosphere and dispersed. Once mixed with air some pollutants persist unaltered and become mixed throughout the atmosphere, thus having a global influence. Most reactive pollutants have a shorter lifetime and are removed either by conversion to normal atmospheric constituents or by deposition on the surface of the earth.

The effects that are of concern are those that do, or may in the future, affect man's health and well-being without undue biological or physical effects. In practice, the association between effects and pollutant concentration is not clearcut because of the immense number of variables involved. This lack of adequate criteria adds to the problem of decision making on the concentration levels acceptable. In general terms, the effects of air pollution are:

1. effects on man, e.g. chronic illness, acute illness, odour;
2. effects on vegetation, e.g. visible, genetic, plant community changes;
3. effects on animals;
4. effects on materials;
5. global changes.

(b) Control of gaseous organic pollutants

Air pollutants, even at their source are generally present in low concentrations, in large volumes of an inert carrier gas. After dispersion in the atmosphere they are further diluted, consequently it is essential that they are controlled before emission at, or as close as possible to, the source.

Removal of pollutants from an emission represents two quite separate problems, depending upon whether the pollutant is gaseous or particulate. For organic gases the following techniques may be adopted: absorption in liquids; adsorption on solids; combustion.

Gas absorption in liquids
Gas absorption in a liquid is a standard chemical engineering unit operation, technically developed and relatively well understood. For high concentrations, a counter-current flow system may be used in a unit such as a packed absorption tower. The collected gases are often stripped from the liquid phase by direct heating and further treated, now as a more concentrated product, by a secondary process.

Adsorption on solids
If the molecules are small then they can be adsorbed on to solids such as silica gel, alumina or charcoal but larger molecules and most organic compounds are best adsorbed onto activated carbon. The activated carbon is then regenerated by heating to 650°C in an inert atmosphere.

Combustion
Combustion involves the treatment of combustible hydrocarbon air pollutants by their complete oxidation to carbon dioxide and water. In some cases a further combustible material must be added to support the combustion. Combustion can be catalytically aided in circumstances where the emission does not contain significant amounts of catalyst 'poisons'. The most common catalysts are platinum, palladium, or transition metal oxides such as cobalt,

chronium and manganese deposited on a support. Catalytic combustion only requires temperatures of 400–500 °C compared with usual combustion at 700–800 °C. Such processes are used widely for the oxidation of organic vapours such as from coffee roasting, paint baking and enamelling.

11.7 CONCLUSIONS

The public is becoming ever more aware of the environmental issues related to the disposal of waste materials, and in particular organic wastes are of major significance. All waste treatment and disposal schemes are in the forefront of media attention, fired by the general public and by pressure groups who are both concerned that, in the past, organic wastes have been disposed of without due regard to public health and safety.

The political scene has also changed, in that politicians of all beliefs have realized the importance of the environmental issues. It is in this field of activity that the most significant changes will take place in the future, since it is only with the agreements of politicians and their servants at both national and regional level that future waste disposal schemes will be given approval.

A further constraint has been imposed by stricter national legislation and international guidelines such as EEC Directives. These will inevitably lead to more stringent discharge standards with the consequence that industry and local government will have to meet the challenge by introducing new and improved techniques. Industry is already aware of this situation and is contributing to these improvements, often by imposing self-regulated quality criteria on its discharges, either as a result of public demand or its own desire to contribute to environmental protection. Frequently, new technologies result from industrial research initiatives which are designed to meet future needs as well as to reduce the financial burden imposed on industry as a result of the requirement to meet stricter standards.

11.7.1 Advantages and disadvantages of degradation

The degradation of organic materials will generally ultimately lead to the production of end-products such as carbon dioxide, water and methane. Such end-products are either stable or capable of acting as an energy source. However, biological degradation also leads to an increase in biomass, which is unstable and consequently requires further treatment and disposal. Whatever are the end-products, industry and the public must be made aware of both sides of the problem.

REFERENCES

Department of the Environment (1983) Special wastes – a technical memorandum providing guidance for their definition, *Waste Management Paper* **23**, London: H.M.S.O.

H.M.S.O. (1973) *Digest of Environmental Protection and Water Statistics* No. 6. London: H.M.S.O.

H.M.S.O. (1978) *River Pollution Survey of England and Wales updated in 1975*. London: H.M.S.O.

OECD (1987) *Compendium of Environmental Data*.

Preston, L.A. (1989) A new horizon for water quality in Japan. *J. Wat. Pollut. Control Fed.* **61**, 578–583.

Part Five

Deposition or Storage

Chairman's introduction

J. Knill

The subject of this session is concerned with the technological implications of the removal of waste from interaction with the environment. Rarely, in fact is it that the title of a session can be, in itself, controversial, but both disposal and storage must be regarded as necessary parts of the waste cycle, and thus the management strategy. They are, in fact, not alternatives. There is, however, a further facet of importance to our discussion in that very long-term storage could become, because of the lack of an acceptable disposal route, a major factor in waste handling options for certain toxic wastes within the third millenium.

Raw waste must be stored for a period before being processed by conditioning, encapsulation and then packaging. A further period of storage may then follow either as a part of the management strategy as is applied for example in the case of heat generating radioactive waste (which requires a period of cooling prior to eventual disposal) or in the situation when there is no available disposal route. Although long term storage has attractions, it nevertheless raises significant problems such as: high construction cost of storage facility often associated with earthquake resistance; lack of appropriately buffered chemical environment needed to avoid deterioration of the waste; exposure of the workforce to toxic doses; requirement for a large number of storage sites; planning difficulties for large surface structures; and uncertainties associated with institutional or environmental change. There is no doubt that we will enter the third millenium with significant volumes of radioactive and toxic wastes in store without any immediate prospect of a disposal route. In some cases storage may be required for some decades before disposal could reasonably be contemplated.

For many years the dilute and disperse philosophy was a catch-all approach to waste disposal. The volume of rivers, lakes or the sea was adequate to ensure considerable dilution, but the pollution of rivers, and eutrophication in lakes and the sea has highlighted that there is a limit to this approach. One must therefore expect to see dilute and disperse being regarded as increasingly unacceptable other than for carefully selected wastes or in particular sets of geological circumstances. That view is likely to become generally adopted, despite well-argued scientific cases for disposal by such means. In those situations where the dispersal takes place within the groundwater system there is a possibility that reaction between the pollutant and the geological environment may result in buffering, and also precipitation of the waste chemical products. Clogging of the geological formation may in addition take place, resulting in improved retention of the liquid waste near to the disposal site.

The particular issue of importance is the extent to which geological containment can isolate wastes for long periods of time. Shallow landfill has historically provided the site for the disposal of all but the most dangerous wastes. The waste will often interact with the groundwater system, the leachate forming a chemical plume within the direction of groundwater flow. As noted previously both attenuation and chemical interactions can reduce the potency, and increase the retention, of the fluids within this pollutant plume. There are considerable attractions geologically in those sites where the water table is relatively deep, thereby reducing the interaction between the waste and an active groundwater flow system. In such cases a cap on top of the landfill site will minimize the influx of surface water, and even climate change in the next millenium may not reduce the integrity of such a disposal site.

The deep disposal option is reserved currently for intermediate and high-level (heat generating) radioactive waste, together with very few toxic wastes; it is possible that the disposal of toxic wastes in such sites will become an increasingly favoured option. The effectiveness of deep disposal relies on the existence of a set of geological and hydrogeological conditions which would effectively isolate the location of the repository from active groundwater flow. Either the groundwater should be stagnant or moving no faster than a fraction of a metre per year. Some geological conditions, such as soluble salt deposits, offer direct evidence of an effectively dry environment. However, construction of the underground chambers, accessed by shafts or tunnels, will disturb the existing conditions and so cause changes such as: the site will act as a groundwater sink; there will be a redistribution of *in situ* rock stresses which may be potentially deleterious; the waste materials may be incompatible with the local geological environment particularly if gas is generated by decay of organic materials; release of organics into the groundwater system can result in change in the chemical retention characteristics of the surrounding rock mass; there may be difficulties in ensuring closure, and sealing at the close of the operational life of the facility, and the re-establishment of the original ground reference condition on which the safety case was probably based may be protracted.

As a result of questions of this type, such a repository may be viewed as a long-term store, a form of half-way house providing monitorable, retrievable storage until safety can be demonstrated. However in such circumstances there will be penalties in that there will need to be a larger volume of excavation, the doses to the workforce may be increased, and there will be an inability to provide for full chemical buffering. Nevertheless, because of the long operational life of a deep waste repository, several decades will pass before closure would be contemplated. Thus there can be a long time to fully evaluate the safety case. One returns almost full circle by this route to the argument for long-term surface storage. Environmental pressure groups argue the case for such storage, often in the form of large, monolithic structures akin to the Mayan Temples or the Pyramids, which would be an obvious warning sign to mankind in the future.

Irrespective of the approach adopted, storage or disposal, the facility must be fail-safe both under present day conditions and in the future when there may be changes in both climate as well as in the institutional arrangements. The longer the period involved the less realistic long-term storage becomes and the better the prospect that safe deep-disposal sites can be found and developed.

12

The assessment to the tenth millennium of the safety of radioactive waste disposal

P.T. McInerney

12.1 INTRODUCTION

Radioactive materials are used, and have been used, for decades in the generation of electricity and in industry, medicine, research and defence. This use results in the production of waste. Regardless of the United Kingdom's future nuclear policy, some radioactive wastes exist and have to be safely managed. There are only two ways to deal with the wastes: these are storage and disposal. A combination of both is required in all cases, but the key objective is to isolate the waste from living creatures.

The nuclear waste can be divided into three broad categories, low, intermediate, and high-level wastes, according to the level of activity per unit weight. The low-level waste arises in the largest amounts and the industry has been disposing of it for 30 years at Drigg, in Cumbria, and in other locations. The intermediate and high-level wastes have been stored. The quantities of intermediate-level wastes are rising to a level which makes a decision about their future storage or disposal important. In the case of the high-level wastes, the quantities are very small and are very radioactive, to the extent that they generate considerable heat. The combination of small volume and heat generation makes cooled storage over 50 years the only practical option.

It can be seen in these introductory facts that storage and disposal each

The Treatment and Handling of Wastes
Edited by A.D. Bradshaw, Sir Richard Southwood and Sir Frederick Warner
Published in 1992 by Chapman & Hall, London, for The Royal Society
UK ISBN 0 412 39390 5, USA ISBN 0 442 31461 2

have very important roles in the safe management of radioactive wastes. Future safe waste management will be achieved by optimizing the combination of the roles, but disposal will be the dominating process, because the long-lived components of the waste must be isolated from living creatures for more than 10000 years. Hence, the important decision is when to move from storage to disposal for each waste stream.

12.2 THE GENERAL STRATEGY

The Nuclear Industry Radioactive Waste Executive (Nirex) was established in 1982 to coordinate, on behalf of the U.K. Nuclear Industry, the development and operation of disposal facilities for low and intermediate-level wastes, within the framework of Government policy. Now U.K. Nirex Ltd, the Company is funded on an agreed basis by its four shareholders: the Central Electricity Generating Board; the South of Scotland Electricity Board; British Nuclear Fuels plc and AEA Technology, with a Golden Share held by the Government. UK Nirex Ltd is required to make its services available for all low and intermediate-level waste arising in the U.K.

To move from waste storage to disposal requires a comprehensive programme of activities, which includes the development and use of a detailed inventory of existing and future wastes arising over the coming decades, a wide range of experimental studies and mathematical modelling to consolidate the present understanding of the behaviours of wastes into the future, field investigations of sites with potential for underground disposal and definition, development and proving of packaging and transport procedures for the waste conditioned for storage and disposal.

National policy decisions about the management of radioactive wastes must be acceptable to the public. Nirex recognises and seeks to respond to public concern by providing the fullest information on its work and plans so that people may be better able to reach their own conclusion.

Nirex's prime responsibility is, therefore, to put forward, on behalf of the U.K. Nuclear Industry, proposals for the safe disposal of low and intermediate wastes, compatible with the national strategy defined by the Government. These proposals will be required to satisfy the requirements of the various regulatory bodies in the U.K., and will be fully assessed by a Public Inquiry.

12.3 THE QUANTITIES OF WASTE

12.3.1 Low-level waste

Low-level waste consists mainly of rubbish, such as discarded protective clothing, used wrapping materials and worn-out or damaged plant and equipment. Since 1959 most of it has been taken to a site operated by British

Nuclear Fuels plc at Drigg in Cumbria. Packaged low-level waste does not require special precautions to protect those handling it, other than rubber gloves, overalls and common sense.

12.3.2 Intermediate-level waste

Intermediate-level waste is about a thousand times more radioactive than the low-level waste. It includes metal fuel 'cans' that originally contained the uranium fuel for nuclear power stations, reactor metalwork, chemical process residues, ion exchange resins and filters. At present, it is stored at producing sites. As the intermediate-level waste emits higher levels of radiation than low-level waste, it is shielded in order to protect people from exposure during storage, transport and disposal. Before disposal it will be retrieved from storage, 'fixed' in a form of concrete and packaged in steel or concrete containers for transport to a disposal centre.

12.3.3 Volumes

Taking into account the continuing use of Drigg for the disposal of low-level waste, it is estimated that Nirex will have to dispose of about 1 million cubic metres of low-level waste and 300000 cubic metres of intermediate-level waste by the year 2030. These figures would approximately double by the end of the lifetime of the repository, i.e. $2.6 \times 10^6 \text{m}^3$ or 4 million tonnes.

To put this in perspective, each year the U.K. disposes of some 30 million tonnes of domestic waste, between 5 million and 10 million tonnes of toxic wastes; coal mining produces some 60 million tonnes of spoil, some of it mildly radioactive; and the china clay industry produces about 20 million tonnes of spoil (RCEP, 1985). Put together, these other wastes would fill the Channel Tunnel every two weeks.

The comparatively small volumes of radioactive waste mean that only one national disposal centre will be needed, which will last for half a century or longer.

12.4 DISPOSAL VERSUS STORAGE

As explained this is not an 'either/or' argument but is essentially a matter of deciding what is the safest combination for each particular waste stream.

So far, in all countries, almost without exception, low-level waste is almost immediately disposed of, and generally this is to shallow land burial sites. Low-level waste requires some 300 years' isolation from man so the storage time of months prior to disposal is not very significant. In the U.K. the main low-level waste disposal facility is at Drigg and the disposal operation there is now carried out to good standards. Capacity exists into the next century. The Nirex commitment therefore is to ensure that there is appropriate additional disposal capacity early in the next century to augment that of Drigg.

Moving to the other extreme of the active waste spectrum, that is the high-level waste, storage is the preferred option for some 50 years. This is necessary because underground disposal would be complicated by the heat which is generated by this highly active material. Additionally, the very small current total volume of this waste, some $1500m^3$ over a 30-year period, means surface storage is, and will continue to be, engineered to a very high standard indeed. This waste is being transferred from cooled storage as a liquid to a solid, vitrified form in a convectively cooled surface store. Ultimately it will require isolation from man for at least 10000 years and so, here again, disposal will be essential.

The intermediate-level waste is the remaining waste stream and to date storage has been adopted, mainly because the quantities, up to the 1980s, had not accumulated to a sufficient level to warrant the development of a large underground disposal project. In 1976 the Royal Commission on Environmental Pollution, chaired by the then Sir Brian Flowers, was critical of the lack of a solution to the safe containment of long-lived highly active radioactive waste for the indefinite future. In 1982, the Nuclear Industry considered it was timely to prepare for an underground repository for low- and intermediate-level waste and this is the main task of Nirex. This is in line with Government policy which was stated in the 1977 White Paper 'Nuclear Power and the Environment' to be to 'secure the disposal of wastes in appropriate ways at appropriate times and in appropriate places'.

Originally, Nirex intended to develop a shallow land burial site to augment the capacity at Drigg, as mentioned earlier. This would have been a safe and acceptable method of isolating low-level and short-lived intermediate-level waste. This was particularly so as the Government later removed short-lived intermediate level waste from the range of wastes permitted in shallow sites. However, as Nirex has to prepare a large, underground cavern for intermediate-level waste, a cost study showed that the co-disposal of low- and intermediate-level waste in a deep repository gave broadly similar costs for low-level waste on a marginal cost basis, as that for fully engineered shallow land disposal. As the co-disposal offered improved isolation this approach was adopted in 1987.

To complete the scene it is necessary to point out that the decommissioning of various nuclear facilities will lead to a category of waste which, although within the low and intermediate range of categories, has the complication of being made up of large individual items. Again, there will be the 'storage through to disposal' process, in that large items may require a period of time for activity to decay to reduce worker dose, before preparations for final disposal are carried out. The option of deep ocean disposal for such items remains a possibility.

For all these waste streams, therefore, there is a need to move from storage to disposal and with isolation periods for intermediate-level wastes of more than 10000 years, the short storage times are a minor component of the total process. For low-level waste, disposal is an established on-going process. The argument, therefore, turns to how soon can safe disposal commence for intermediate waste?

Some of the basic reasons why this type of waste should be disposed of relatively early rather than kept indefinitely in a surface store are listed below.

1. On-site storage would require numerous small sites, all of which will have to be looked after, rather than one single disposal centre upon which resources could be concentrated. Indefinite storage would also prevent the sites being returned to a 'green field' state. Sites of existing nuclear installations were chosen for reasons other than suitability to contain radioactive waste, although some have potential.
2. Accidental or intentional damage is a greater risk at many surface stores than at a single centre deep underground. Any disruption to surface buildings might release radioactive material directly into the human environment, whereas a disposal centre deep underground will provide a geological and physical barrier of at least 200 m.
3. The operation of surface stores will mean people being exposed to radiation during maintenance work, repackaging and ultimately during the rebuilding of the stores.
4. Any policy for continued storage rather than for a custom-built disposal centre is simply putting off a decision that will have to be taken one day. As the technical ability to dispose of these wastes exists now, a storage policy is doubly suspect.

Nirex is therefore working to build one national deep disposal centre for low and intermediate-level wastes to serve the nation for at least 50 years.

The following sections give an overview of the comprehensive programme of Nirex directed to establishing the case for a safe, deep underground repository.

12.5 REQUIRED SAFETY STANDARDS

The Government has laid down stringent safety targets for the selection, design, construction and operation of a disposal centre for radioactive waste.

There is nothing new or unique about radioactivity. Humankind has evolved in a naturally radioactive environment, subject to radiation from the Earth, from outer space, from within the human body, even from cultivating the ground and living in buildings. This natural background radiation accounts for nearly 90% of the total amount of radiation which the average person in Britain receives in the course of a year. Most of the balance is from the medical use of X-rays and radiotherapy. Members of the public have no concern about changing their radiation dose by high percentages by moving between different parts of the Country with very different natural background radiation levels.

A disposal centre will be subject to authorization under the Radioactive Substances Act 1960. It is a government regulation that all radiation exposure should be as low as reasonably achievable and that the radiation exposure of the most exposed member of the public should not exceed a specified target at any time in the future. The target value for a disposal centre for radioactive waste is equivalent to a radiation dose of 0.1 millisieverts (mSv) per year. It is about one third of the dose each person gets in a year from naturally occurring radioactive material in their bodies, one twentieth of natural background radiation from the Earth's rocks, the Sun and outer space, and one two-hundredth of the dose people living in some houses in granitic parts of the Country, such as Cornwall, get from their surroundings. In the case of the disposal centre this peak dose, if it comes at all, would be in the period beyond 10000 years. Nirex will have to show at a Public Inquiry that in disposing of radioactive waste this target can be achieved.

12.6 RESEARCH AND DEVELOPMENT

The basic model used to represent this underground repository in the R&D programme is one comprising a multi-barrier containment approach. Disposal of radioactive waste underground removes the direct radiation hazard completely. The second objective is the elimination of the remaining risk of radioactive particles entering the food chain or drinking water or being dispersed in the air in harmful quantities. Hence, the research and development programme must provide data for the component parts of the multi-barrier model and for the emplaced waste, and these scientific data are used in mathematical models which predict the long-term evolution of the repository.

The internationally accepted way to achieve this is to process the waste into solid forms and to place them in an underground disposal centre. The so-called 'multi-barrier approach' (Fig. 12.1) is designed to keep the radioactive substances away from people, animals and plants with a series of different and separate physical and chemical barriers until the process of radioactive decay makes them barely distinguishable from naturally occurring materials. Indeed most of the long-lived waste contains naturally occurring materials like uranium. The 'multi-barrier approach' sets standards for the safe disposal of radioactive waste which go far beyond those currently used in the United Kingdom for the disposal of other toxic and hazardous wastes.

The aims of the research programme are threefold.

1. It must provide the data necessary to carry out assessments. These include solubilities and other chemical parameters of radionuclides, physical and chemical properties of geological materials, the biological and

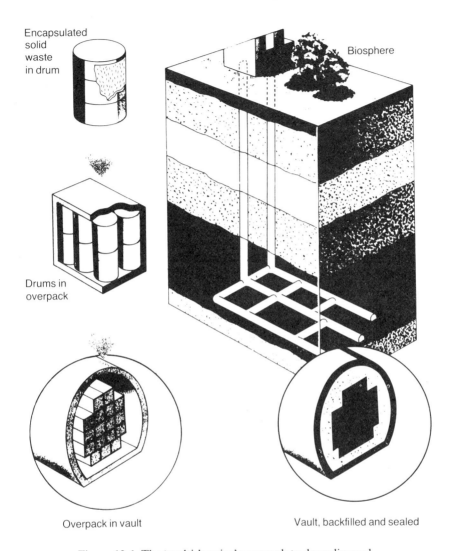

Encapsulated
solid
waste
in drum

Biosphere

Drums in
overpack

Overpack in vault

Vault, backfilled and sealed

Figure 12.1 The 'multi-barrier' approach to deep disposal.

microbiological conditions of the location, etc. It is necessary to provide estimates of the uncertainties in such data as well, so that these too can be incorporated in the assessment. The research also provides mathematical models that describe the processes occurring in a waste repository and its environs; these are used to predict the evolution of these processes in time.

2. Knowledge of the processes that limit the release of radionuclides from a repository enables informed choices to be made concerning repository

materials and design. Thus advice is obtained from the programme on suitable backfill or container materials, host geologies, etc. and hence helps optimize the repository design and choice of location.

3. An important aspect of the research is to examine the key assumptions underlying any assessment. These assumptions include such things as the durability of the repository components, the behaviour of groundwater, the migration of gas through the repository and the evolution of the biosphere. The assessment models are necessarily simplified but, since they are based on this detailed and scientific framework, we can be confident of their reliability. The robustness of the final safety case is also enhanced by participation in international collaborations, both on scientific questions of general importance and on computer code validation and verification.

The natural pathway for the radioactivity contained within the repository to travel back to Man is via the groundwater that permeates the deep rock strata. The immense delays involved in such a path work in favour of protection as radioactive toxics have a finite lifetime and containment for a sufficient length of time will render them harmless. It is essential to show that the inevitable physical and chemical processes do not lead to any significant risk. Thus much of the research programme is concerned with analysing the natural evolution of the repository and has concentrated on the groundwater and related gaseous pathway. Of course, the assessment must consider other possible routes including intrusion by people into the repository or natural disruptive events.

The natural evolution of the repository can therefore be imagined in the following way. The waste will be contained within steel and concrete containers and the free space within the repository will be filled with concrete. In the years immediately following backfilling with this concrete, water will permeate and saturate the region. As it equilibrates with the concrete it will reach a high alkalinity, which will be of value in reducing both the steel corrosion and the solubility of the radionuclides. The steel of the containers will slowly corrode and eventually allow radionuclides to dissolve in the water of the near field. Many of the radionuclides have very low solubilities. The concentrations in solution will be further reduced by their tendency to stick to the surfaces of the pores in the concrete. Thus the repository chemistry will determine the eventual source of radionuclides that leach from the repository. Radioactivity will migrate from the repository in the very slow groundwater flow of the chosen site. It may take hundreds of thousands of years for water to travel from depth up to the surface. However, most radionuclides will travel much more slowly than the groundwater due to their tendency to stick or sorb onto the rocks through which they must travel. When they reach the biosphere, the nuclides travel through the unsaturated soil zone to be taken up by plants and to disperse in rivers and seas. However, it is predicted that these processes will occur so long in the future that much of the activity will have naturally decayed away. All these processes are examined in detail and quantified by the research programme.

In addition, the consequences of gas generation must be assessed. Gas is produced by the corrosion of the steel and the degradation of the waste itself. The escape of this gas may provide a route for radionuclides to reach the surface other than via the groundwater.

When the radionuclides finally reach the biosphere, it is likely to be at immensely long times into the future. Thus, to assess properly the consequences, we must consider the evolution of the climate and land over these timespans (Fig. 12.2).

A deep repository is well isolated from extreme events on the surface. However, long-term changes in the land surface arising from fluvial or glacial action could modify the groundwater flow pattern or disturb a repository that was not sufficiently deep.

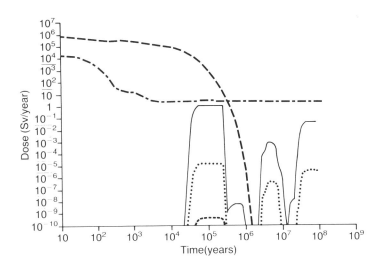

Figure 12.2 An example of a calculation for the groundwater pathway. The roles played by the repository materials, the geology and biosphere are illustrated by plotting against time the doses corresponding to the near-field porewater concentration both neglecting and including near-field chemical effects (– – – – – and – – – –, respectively), the dose corresponding to the water emerging from the geosphere (——), and the maximum individual dose received from both terrestrial (· · · · · · · ·) and marine biosphere (- - - -). The doses corresponding to concentrations are evaluated by assuming a hypothetical individual takes all his drinking water from the porewater. This would, of course, not be practical.

Different pathways and processes are expected to be important at different times in the future. It is, therefore, helpful to think of the assessment in terms of a number of different time-frames. Five natural time-frames are currently considered; 0–300 years, 300 to 10^4 years, 10^4–10^5 years, 10^5–10^6 years and 10^6–10^8 years. The current interglacial conditions are expected to persist over most of the first two time-frames, although a major uncertainty in the climate arises from the greenhouse effect. Glacial–interglacial cycles are then expected over the next two time-frames. Substantial variations in climate and sea-level will be associated with these. Greater climate stability is a possibility over the final time frame. Although general tectonic stability should be preserved, limited tectonic changes could occur. As a consequence, there is considerable uncertainty in assessments over this time-frame.

These natural time-frames should be contrasted with human experience, where 10^3 years (representing 30–50 generations) is regarded as long-term. Therefore, assessments over long time-frames might be less detailed than those to, say, 10^4 years.

The effects of gas generation are of most concern over the first two time-frames, when most of it is generated. Restrictions on the use of the site might be expected to end early in the second time-frame. The consequences of intrusion have to be addressed after this time, and they are expected to be most important over the second and third time-frames. The first radio-nuclides are not expected to return with the groundwater until time-frame 3 or 4. So the groundwater pathway is most important for the last three time-frames.

12.7 SITE SELECTION

In searching for sites the International Atomic Energy Agency (IAEA) suggests a three-stage process which Nirex is following. This comprises:

1. National survey and evaluation: this defines areas having favourable characteristics for a disposal centre based principally on geology and hydrogeology.
2. Site identification: the areas defined in stage (1) are scanned for available sites and desk studies are carried out to assess their suitability so that a small number of the most suitable sites can be investigated.
3. Site confirmation: those sites identified in stage (2) are investigated thoroughly by geological and geophysical means to confirm the desk studies.

Nirex has completed the stages (1) and (2), and is now proceeding to stage (3). This will include a comparative assessment of the investigated areas leading to the selection of one site for development, if it is shown that a disposal centre could be safely built there.

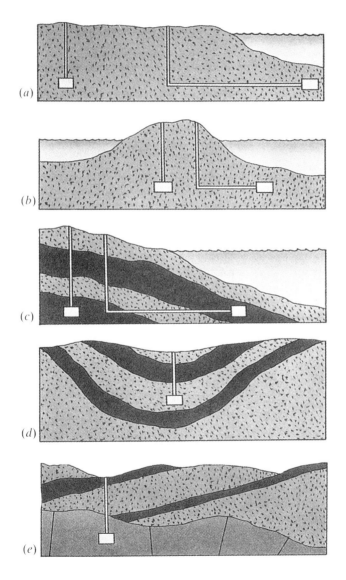

Figure 12.3 Suitable geological environments for a deep repository: (*a*) hard rocks in low relief terrain; (*b*) small islands; (*c*) seaward dipping and offshore sediments; (*d*) inland basin; (*e*) basement rocks under sedimentary cover.

12.8 WORK ON AREAS OF SEARCH

In looking for a site, considerations of safety are paramount. For reasons already explained, a deep disposal centre is best placed in a formation with very little groundwater movement. Accordingly Nirex defined five potentially suitable environments (Chapman *et al.*, 1986). They were as in Fig. 12.3.

1. Hardrocks in low-relief terrain: low-relief areas have little driving potential for groundwater flow. What there is tends to be controlled by major fractures within the hard rock.
2. Small islands: an island sufficiently far from the coast will have its own groundwater flow system, independent of the mainland. Beneath the island's seawater–freshwater interface there may be very slow-moving groundwater.
3. Seaward dipping and offshore sediments: in this environment, groundwater flow is expected to be very slow and inclined towards the sea. Offshore the groundwater will tend to move very slowly upward and, over geological timescales (i.e. millions of years), eventually into the sea.
4. Inland basins: these are deep basins of mixed sedimentary rocks. Groundwater flow is controlled by those formations of higher permeability and tends to dip towards the centre of the basin. Flow in the lower permeability sediments is very slow and tends to be vertically upwards.
5. Low permeability basement rocks under sedimentary cover: environments occur where the groundwater flow is predominantly in the sedimentary cover with little anticipated connection to the basement rock.

The initial two sites to be investigated are at Sellafield in Cumbria and Dounreay in Caithness.

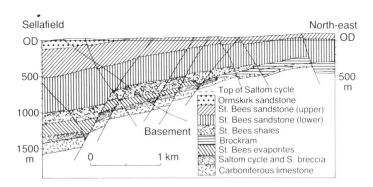

Figure 12.4 Predicted cross-section of the geology in a northeasterly direction from Sellafield.

12.9 SELLAFIELD, CUMBRIA

One of the main reasons for selecting Sellafield as a potential location is the fact that about half of Britain's radioactive waste is either stored or produced at the BNFL site. Additionally a repository would be compatible with such a large nuclear site and hence have less environmental impact. As a result transport costs would be kept to a minimum, and very little would be required in the way of additional transport access. The Sellafield site already has adequate road access from the A595. Rail access could be made from the existing spur, or by providing a new link from the Cumbrian coast line.

The traditional industries of Cumbria are mining and heavy industry, and these skills, together with the local nuclear experience could be valuable in a programme of disposal centre construction and operation.

The local geology is more complex than at the other potential location (Fig. 12.4) but there is confidence that the groundwater movement can be successfully calculated and predicted. The formation of interest for disposal centre construction is the Borrowdale Volcanic series, which is a hard rock.

12.10 DOUNREAY

It is believed that the UKAEA site at Dounreay offers potentially suitable hard rocks for the development of a disposal centre. If a centre were constructed it would be in the hard rock of the Moine metasediments or the Reay Diorite. The geology of this area (Fig. 12.5) is predictable and there is confidence that the groundwater flow can be modelled. Water movement in these rocks is expected to be slow because of the flatness of the landscape. The relatively simple geology increases confidence in long-term predictions and also allows some flexibility in the choice of disposal centre design.

In the wider region there is a large workforce with nuclear-related experience. Depending on the future of the Dounreay establishment some of this experience may be available to Nirex, during the construction and operation stages of the waste centre.

12.11 TRANSPORT AND CONTAINERS

Work is being undertaken to establish the transport system for: a range of low and intermediate-level packages; construction materials for the disposal centre; spoil from the disposal centre excavation, and the workforce to and from the disposal centre. Road, rail and sea are being considered individually and in combination as appropriate. The feasibility, costs, impact on the environment, construction of new transport facilities and operation of the transport system are all being addressed. In addition the equipment needed to package and transport the waste is being developed (Fig. 12.6).

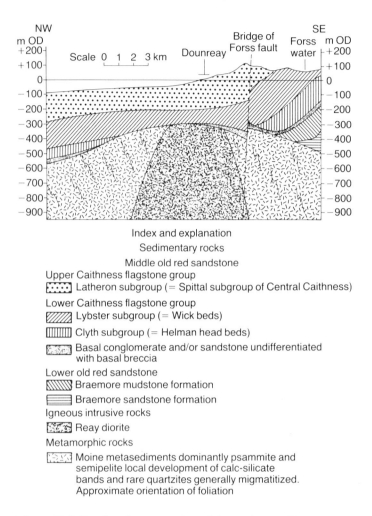

Figure 12.5 Predicted cross-section of the geology at Dounreay.

12.12 NON-NUCLEAR ENVIRONMENTAL ASSESSMENT

Preliminary assessment work of the non-nuclear environmental impact of disposal centre development continues to be directed at the areas identified for investigation. Careful consideration is being given to ways in which the impact of a large development such as this can be minimized. An Environmental Assessment will be carried out in accordance with the guidance given by Government departments to comply with the EEC directive.

Nirex Standard Waste Containers			
For Low-Level Wastes	*Description*	*Application*	*Max. external Dimensions (m)*
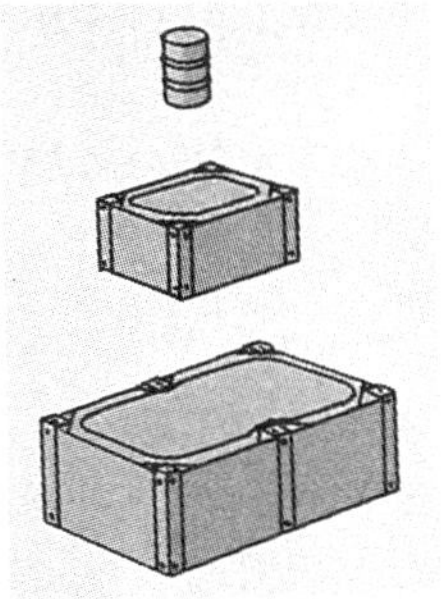	200 litre drum	The normal container for most operational wastes	0·61 OD × 0·863
	3m³ LLW Box	For operational and decommissioning wastes unsuitable for 200 litre drums	2·15 × 1·5 × 1·3
	12m³ LLW Box	For large items of Decommissioning wastes	4·0 × 2·4 × 1·85
For Intermediate-Level Wastes	*Description*	*Application*	*Max. external Dimensions (m)*
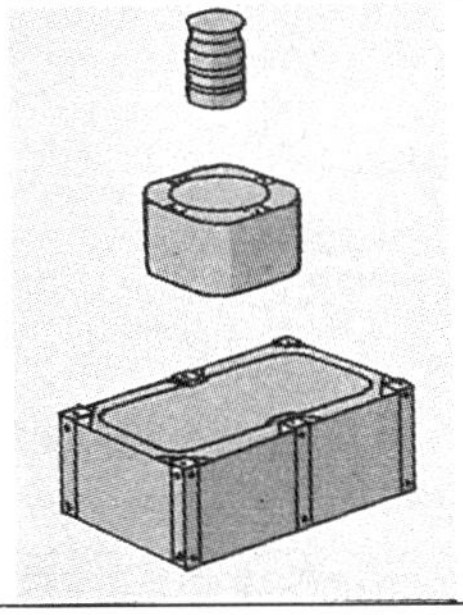	500 litre drum	The normal container for most ILW.	0·8 OD × 1·2
	3m³ ILW Box	For wastes unsuitable for 500 litre drums	1·72 × 1·72 × 1·2
	12m³ concrete box	For large items of decommissioning wastes	4·0 × 2·4 × 1·85

1. The dimensions are provisional. 2. Drums and boxes will be manufactured in steel apart from the 12m³ concrete box. 3. 1m³ = 1 cubic metre = 1000 litres.

Figure 12.6 Nirex standard waste containers.

12.13 REPOSITORY DESIGN CONCEPT

A depth of at least 200 m is judged desirable cover for the disposal of long-lived intermediate-level waste. The depth is chosen to provide a long potential pathway for the migration of any radioactive substances to the environment and adequate protection during the next Ice Age, some 20000–30000 years hence. The actual operating depth will be dependent upon the chosen geological environment and the site itself. The depth will be chosen after detailed field studies, and an underground investigation will be conducted so as to confirm the groundwater conditions.

At such depths, several engineering concepts are potentially suitable. A disposal centre could be situated either under land or beneath the seabed; caverns, tunnels or boreholes could be used for disposal. Nirex considered and evaluated three concepts: under land, accessed from a land base; under the seabed, with access from the coast; under the seabed, with access from an offshore structure. Existing caverns, such as mines, will have a documented history and well-understood geology, although there are few in the U.K. that show any great potential for the disposal of radioactive waste. They may be unstable, in areas of relatively fast-moving groundwater, or be liable to gas problems.

The construction and operation of a disposal centre deep underground will use well-established technology. A shaft to the required level of rocks will be constructed, and then caverns or tunnels excavated and supported, if necessary, to form disposal vaults.

After the waste packages have been stacked in the vaults, backfilling with suitable grouting material could be carried out. This would provide long-term physical stability to the vault and also establish a further barrier to slow the movement of radioactive substances.

Services required at a disposal centre include transport facilities, ventilation, power generators, interim storage and, of course, safety facilities. Additionally, areas for monitoring and decontaminating transport vehicles, if necessary; handling any non-standard waste packages; a concrete preparation plant; and all the offices and parking areas associated with modern industrial operations, will also be required.

Feasibility studies have been carried out to consider excavation in the various suitable geological formations of hard and soft rock, to gain an understanding of the costs and constraints of excavation.

All of these studies suggest that disposal vaults of large cross-section will be more efficient in terms of space utilization than small tunnels. This would tend to favour hard rock, such as anhydrite or granites, rather than soft rock such as mudstone or clay.

It is not expected that anyone should ever want to retrieve waste from the disposal centre, but the ability to do so could be retained. The waste will be monitored for as long as it is felt desirable.

As a result of these studies there is no doubt that an underground disposal centre can be safely constructed. Further studies involving the development of these designs have been undertaken. A conceptual design for a land-based disposal centre is shown in Fig. 12.7.

12.14 PUBLIC DISCUSSION OF THE ISSUE

Following cessation of the near-surface investigations in May 1987, the Company carefully considered the most appropriate means of developing proposals for a combined deep disposal centre for low and intermediate-level wastes. Recognizing the importance of striving for public acceptability, providing that technical excellence and safety are not compromised, the Company launched an initiative in public awareness and discussion.

In planning a deep disposal facility, the Company had identified the range

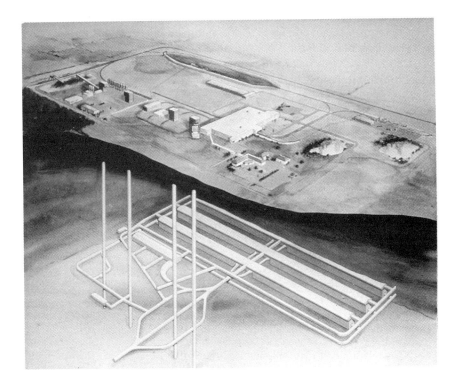

Figure 12.7 A conceptual design for a land-based disposal centre.

of design options and areas with geological potential and wanted to stimulate comment at that very preliminary stage to assist the Company in developing acceptable proposals. Acknowledging the importance of understanding the broader social context in which the Company needs to carry out its technical responsibilities, a decision was taken to prepare a comprehensive discussion document. This publication, entitled *The way forward*, was intended for a wide audience. Considerable effort was put into producing a simple but informative document, not too technical in content but addressing a range of issues which are predominantly technical in origin.

The way forward, on which this paper is based, was launched in November 1987, at a meeting in London attended by Members of both Houses of Parliament, and of the European Parliament (UK Nirex Ltd, 1987). The launch provoked considerable new media interest which itself brought widespread public interest in Nirex's Discussion Programme.

Altogether some 50 000 copies of the full document and its summary were distributed during the discussion period which lasted until June 1988.

This massive exercise elicited more than 2500 responses. These were passed to the Environmental Risk Assessment Unit (ERAU) of the University of East Anglia which was contracted to provide a full independent analysis of the responses received. The ERAU published their report in November 1988 (Environmental Risk Assessment Unit 1988).

Additionally, Nirex has a wide range of information services to ensure the fullest information on its work is available to the public.

12.15 CONCLUSION

The safe management of radioactive waste is a challenge that the nuclear industry is undertaking in a responsible and comprehensive manner. Its proposals will be subjected to thorough regulatory assessments and comprehensive public debate. The industry is determined to put in place facilities which will ensure adequate safety for radioactive waste arisings to beyond the tenth millennium.

REFERENCES

N.A. Chapman, T.J. McEwen and H. Beale (1986) *Geological Environments for Deep Disposal of Intermediate-Level Wastes in the United Kingdom.* IAEA-SM-289/37.

Environmental Risk Assessment Unit (1988) *Responses to The way forward.* University of East Anglia.

Royal Commission on Environmental Pollution (1985) *Evidence to Royal Commission on environmental pollution, 11th Report, Managing waste the duty of care.* CMD 9675. London: H.M.S.O.

UK Nirex Ltd (1987) *The way forward. A discussion document.* November 1987. Didcot: UK Nirex Ltd.

13

A perspective from the waste management industry

H.G. Pullen

13.1 INTRODUCTION

There is no question that pressure for change exists. Society has moved from a position of accepting the supposed inevitable consequences of industrial activity, as a price which has to be paid for the material benefits, to a position where these consequences are now actively being challenged. Challenge and informed discussion should generate considered change. Earlier excuses for lack of action, such as threats to livelihood or cost, are no longer acceptable and will have little credence in the 1990s and beyond.

Pressure for change is not only confined to those who generate waste as a by-product of a manufacturing process, it is also to be found within the waste management industry, which has the task of either applying treatment technologies to these wastes, where appropriate, or becoming their long-term custodian. The waste management industry, primarily exists to manage and engineer waste streams for which treatment technologies do not exist or are uneconomic to apply to the quantities involved.

13.2 CHANGE IN THE WASTE MANAGEMENT INDUSTRY

Anticipating the nature and scope of change in most industries, particularly over a ten-year period and beyond, is fraught with problems. Changes in waste management are probably even more difficult to predict, as the pressures that influence the business are diverse, often driven by public

The Treatment and Handling of Wastes
Edited by A.D. Bradshaw, Sir Richard Southwood and Sir Frederick Warner
Published in 1992 by Chapman & Hall, London, for The Royal Society
UK ISBN 0 412 39390 5, USA ISBN 0 442 31461 2

misconception, or naive environmental reasoning, or without adequate scientific justification. Thus taking a view into the 21st century presents a major challenge.

Within the waste management industry, the response to change has hitherto been slow. Report upon report has commented and warned of some of the areas of risk within this industry and recommended that action should be taken. However, changes within most industries are inevitably slow unless there exists either a tangible benefit, usually economic in origin, or external pressure resulting in legislative changes.

The waste management industry, that is producers and public and private sector contractors, is no different. It has generally responded and amended its practices as a consequence of legislative changes. Manufacturing industry has generally been prepared to accept the status quo on the basis that if it did not, then its disposal costs would rise.

Fortunately, a number of major waste management companies have recognized that there are sound commercial benefits to be derived by taking the initiative and investing, often alone, in technologies in advance of external pressures. Waste treatment, incineration and recycling plants are typical examples. Currently, however, such technologies are still only applied to less than 10% of the U.K.'s hazardous waste arisings and to less than 1% of all Controlled wastes (Control of Pollution Act, 1974). Disposal by landfill is the predominant technique for the largest proportion of wastes. Thus whilst there will be a gradual increase in the application of alternative technologies to the treatment of wastes, and a progressive movement of wastes away from the landfill disposal route to these technologies, landfill is likely to remain the disposal route for the major part of the Controlled wastes well into the 21st century.

Landfill of wastes has been employed by most urban societies for hundreds of years and has progressively changed from burning rodent-infested dumps to, in many cases, well equipped professionally operated facilities. Regrettably not all such public and private sector facilities currently meet acceptable standards and, without properly applied enforced legislative pressure, are unlikely to change.

13.3 LEGISLATIVE CHANGES

The 1990 Environmental Protection Bill (EPB) currently before Parliament attempts to address this difficult question of standards and provides the framework for the creation of a regulatory structure which, if properly resourced, should progressively bring about the improvements that are sought (now the Environment Protection Act 1990).

13.3.1 The major elements of the bill

(a) Part 1 – integrated pollution control (IPC)

The revision of central pollution control in the U.K. will result in the major polluting processes (amounting to some 3300 operating processes) being required to use the 'best available techniques not entailing excessive costs' (BATNEEC) in relation to these processes. This is to be regulated on an integrated basis by reference to the relative effect on air, water and land by the application of the principle of 'best practicable environmental option' or 'best practicable means' according to different definitions. The Government has issued for consultation, details of those processes and substances to be controlled by IPC. Waste disposal plants, i.e. incineration and waste treatment have been indicated as candidates for inclusion within IPC, thus removing them from local authority control. Control will be exercised by Her Majesty's Inspectorate of Pollution (HMIP) with the use of considerably enhanced powers to issue enforcement, variation and prohibition notices to control the processes.

(b) Part 2 – waste on land

1. The creation of single role Waste Regulation Authorities (WRAs) by the separation of the regulation from the operation by the formation of Local Authority Waste Disposal companies (LAWDCs).
2. An extension of the powers of a WRA to revoke or refuse the transfer of a waste management licence where the applicant has relevant convictions; or has insufficient resources to assure meeting licence conditions; or lacks technical competence.
3. Waste management licences to remain in force after the completion of the disposal facility until the licensing authority issues a certificate that the facility is safe and no longer a threat to the environment HMIP's view in 1990 is on a 50-year timescale.
4. To make the breach of a licence condition an offence.
5. Waste disposal site operators to be charged for licences and the inspection and monitoring by the licensing authorities. These charges will be significant and not just based on administrative costs.

(c) Duty of care

In Clause 26 of the EPB the Government has set out its proposals for a Duty of Care to be applied to all those who import, produce, carry, keep, treat or dispose of controlled wastes or as brokers have control of such wastes.

The Duty of Care will involve producers and other holders of waste in shared responsibility for ensuring that Controlled waste is not illegally managed, and that it does not escape from control and that it is transferred only to an authorized person and that it is adequately described to enable proper handling and treatment. Breach of the Duty will be an offence.

The duty of holders is limited to taking such measures as are reasonable in their particular circumstances. The Duty of Care does not necessarily apply strict as opposed to absolute liability on waste producers. The Government does not consider it reasonable to hold a person liable for defaults by subsequent holders of the waste who are outside of their control. The question to the courts will be the meaning of the word outside. The Duty will apply to all holders of the waste; each person in the chain will be responsible for his own actions.

(d) Special wastes

In Clause 52 of the Bill the Government has retained the Special waste powers of cradle to grave tracking of wastes through a consignment note system. The long awaited review of the scope of the Special waste regulations has recently been issued as a consultative document and contains a number of new proposals of significance.

1. A new definition of special wastes which takes account of the EC draft Directive on hazardous wastes, the definition of such wastes in the Basle convention and the work undertaken by the OECD on the classification of waste. It is proposed that Special waste will be defined by reference to eleven characteristic properties which cause waste to be dangerous or difficult to dispose of. The new definition includes substances which present a risk to the environment as well as those harmful to human health.
2. The consignment note system is to be strengthened and improved. The consignment note will itself be revised to provide a precise description of the waste and will require those involved to provide an explicit certification about the waste, its transfer and disposal.
3. There will be a requirement that before Special waste is removed from its point of generation or storage, the holder must enter into a contract with the disposer to ensure that disposal takes place on an appropriately licensed site.
4. Regulatory Authorities will be required to recover a reasonable charge for the provision of consignment note forms, as well as the cost to the authorities of administering and monitoring the operation of the consignment note system.

The foregoing changes to waste management regulation within the U.K. are therefore likely to have a significant influence on practices and procedures in the 21st century. They will only achieve their aim if adequately resourced in terms of the number, quality and motivation of those working in the regulatory and inspection fields.

13.4 EUROPEAN LEGISLATION

We have already seen a significant amount of legislation derived from the European Community built into the U.K. waste management controls, and this process will continue and is likely to be a dominant feature for change in the future.

While the U.K. Government can currently block a draft directive believed to be unacceptable, by the mid 1990s draft environmental directives will require approval by only a majority of ministers before they are adopted. This will have major implications on waste management within the U.K. if directives are approved which do not adequately reflect the circumstances of, for instance, regional geology, current waste industry structure or existing national regulations and controls.

Clearly, the rate at which specific waste topics are addressed within the EC and draft proposals produced will have an influence upon the rate of change within the U.K. waste management industry, but there is firm evidence that their rate of production is accelerating often at the expense of their precision in wording.

13.5 RECYCLING AND WASTE MINIMIZATION

Notwithstanding the pressure for change within waste management as a consequence of legislation, it is also quite clear that the issues of recycling and waste minimization will force changes, particularly in the quantity and nature of wastes for disposal. U.K. Government proposals, EC initiatives and environmental pressure groups are likely to be strong influences in this area. Recycling initiatives are increasing and within the next 10 years are likely to have a real impact upon wastes, particularly in the domestic area. Segregation of waste by householders will inevitably change the composition of domestic waste with, however, little reduction in the pollution potential of that which remains. Recycling activities within industry often have the consequence of producing a waste of a higher concentration, albeit in a much reduced volume than hitherto. This can, in some instances, create problems with its final handling and disposal.

Waste minimization within manufacturing units and at the design stage of new products and processes will also lead to a reduction in the overall quantity of waste available for disposal in the future. Over the last 5 years many organizations have introduced such schemes and the consequneces are already to be seen in waste volumes.

13.6 PUBLIC PRESSURES

The waste management industry is not unique in that its development has often been inhibited by its failure to obtain public confidence in its systems and practices. To develop treatment and disposal facilities it requires permits and licences sought through the democratic processes, yet in many instances it has failed to explain to the general public, in a manner they are able to understand, why the facility is necessary and how it is to perform. Without this public awareness and confidence, obtained by a disclosure of relevant information, and adopting open door policies, the public's attitude will continue to be negative and hostile. An unfortunate consequence of this failure to inform could be that sophisticated higher technology solutions to the management of waste will be demanded and given public approval without adequate debate as to their necessity or likely success.

The on-plant treatment of waste streams, as part of a recovery or recycling process, is often feasible where contaminants are amendable to separation, is widely practised and is likely to grow, particularly as the costs of off-site treatment by the waste management industry increase.

13.7 CHANGES IN WASTE TREATMENT PRACTICES

A fundamental truism concerning waste is that it is not produced to a tight specification. Variations in waste compositions occur for a variety of reasons ranging from process fluctuations to differing products actually being produced by the same plant in batch processes. Such variations are compounded when the historical development of a manufacturing site has made no provision for the segregation of process waste streams. The variability of waste is a major inhibition to the application of many technologies to the treatment of waste streams.

It therefore follows that any treatment system for wastes should be capable of accepting these variations and continue to operate to the anticipated standards.

13.8 INCINERATION

High temperature incineration of combustible wastes has an increasing role to play in this area and is a good example of a high technology treatment system, able to accommodate significant variations in the waste composition. Merchant incinerators operated by the waste management industry demand for their success the ability to blend wastes from differing sources to provide a consistent and uniform feedstock of the required characteristics. On the other hand, many in-house incineration plants do not have the ability to call upon a wide range of waste sources, and any significant change in the production processes can have an adverse affect upon the plant's ability to continue to operate. This applies particularly where a high calorific value stream is removed and upon which the plant is dependent. In-house disposal by incineration is therefore not always a satisfactory option where variations in the feed can be difficult to accommodate.

13.9 CO-DISPOSAL LANDFILL

Considerable flexibility and the ability to accept substantial feedstock variations is also found in co-disposal landfill where industrial and domestic wastes are disposed of in the same site. The U.K. Government actively supports this method of disposal, although some countries have either banned it or heavily restricted its use. The origins for the lack of support for this technique are varied but are largely based upon unfortunate pollution incidents due to a lack of understanding of the requirements of co-disposal and poor selection in inappropriate geological systems.

Increasingly, co-disposal landfill is being described in terms of being a bioreactor, where operating parameters such as the hydraulic retention time within the reactor are measured in years. A substantial body of evidence is now being assembled which confirms that within the intensively active biological regime within the bioreactor anaerobic and aerobic processes break down and oxidize complex organic substances. In addition, the mobility of metals initially present in either the domestic waste or the industrial waste are very much reduced through a range of mechanisms including precipitation as hydroxides carbonates or adsorption on to other organic or inorganic substrates. The potential of domestic refuse to neutralise acid and alkali wastes has also been recognized as a valuable feature in this bioreactor.

A co-disposal reactor therefore differs substantially from the so-called secure landfills which in many instances are nothing more than secure heavily engineered repositories for wastes with little or no possibility of a reduction in pollution potential from the wastes occurring with time. The fact that they are so expensive to construct makes them uneconomic for the disposal of domestic waste an essential component of the feed to the bioreactor.

Studies have shown that a bioreactor co-disposal landfill located in a suitable geological formation which prevents contamination of groundwater, or engineered using natural or synthetic materials to achieve the required degree of isolation, when properly managed can eliminate the potential hazards of the deposited wastes.

A feature of this technique which should not be under-estimated is that it is largely immune to variations, within limits, in the composition of the waste feed. The buffering capacity of a system with a long hydraulic retention time is considerable. It is thus ideally suited to the treatment of a wide range of variable wastes: a feature that no engineered process other than high temperature incineration could possible tolerate.

Because the technique depends upon the activity of the microbial population within the reactor, careful monitoring of the process and selection of the industrial wastes is however important if maximum efficiency is to be maintained. For instance, excessive quantities of a high concentration of bactericidal materials would have a disastrous effect; whereas limited quantities of moderate concentrations would be tolerated.

Recognizing this need for feed consistency in any microbial system, and to provide controlled conditions, there are plans to construct a liquid waste pre-treatment plant at a major co-disposal site in Essex, with the object of reducing the variations in the feed composition. To achieve this incoming wastes are to be treated to establish a neutral pH regime, selected agro-chemical waste will be subjected to acid or alkali hydrolysis, and specific pre-treatment applied to other wastes before the total liquid feed after balancing is injected into the domestic refuse deposits. In this way the bioreactor will be provided with a balanced, neutral feed to encourage the rapid biological degradation of the components of the liquid wastes. This biodegradation could not be provided outside of the confines of a co-disposal bioreactor.

Although the foregoing largely reflects the existing situations within the U.K., it has to be recognized that alternative engineered treatment technologies for many waste types will be proposed. In some cases these will be a total treatment, particularly where recyclable materials are involved, where there will be little if any in the way of residuals to be dealt with elsewhere. However, for most of the following processes there will remain a need for the disposal of residuals in a smaller but in a more concentrated form. In addition, application of these processes will in some cases only be possible after pretreatment of the waste prior to final treatment.

The most common types of treatment alternatives are:

1. chemical treatment: neutralization, precipitation, ion exchange, solidification/fixation and dechlorination;
2. physical treatment: Solid/liquid separation processes (screening, sedimentation, flotation, filtration and centrifugation) membrane separation processes (dialysis, reverse osmosis, ultra-filtration, and electro dialysis) evaporation, distillation/steam stripping, solvent extraction and adsorption;
3. biological treatment: activated sludge, trickling filter, aerated lagoons, waste stabilization ponds, and anaerobic digestion;
4. thermal treatment: calcination, pyrolysis, open burning and incineration (rotary kiln, fluidized bed, multiple hearth).

13.10 CONCLUSIONS

The 1990s and beyond will be an era of change in waste management, largely brought about by national and European Community legislation and pressure from environmental groups. These changes will result in more waste recycling and active waste minimization programmes, coupled with the introduction of more readily disposed of products. Many recycling initiatives will, however, demand economic support for their success.

Pre-treatment of waste on site or as part of a pre-treatment process prior to disposal to a secure landfill bioreactor are likely to be more common, as is the application of high temperature incineration to the disposal of a wider range of wastes.

Waste management has undergone substantial change in the last decade, and given the appropriate level of staffing with the greater professionalism needed in the regulatory and operational fields, these changes will continue and the U.K. will enjoy the benefits of a high standard, technically sound waste management industry without the environmental weaknesses found in other parts of the world.

REFERENCES

Control of Pollution Act (1974). London: H.M.S.O.
Environmental Protection Act (1990). London: H.M.S.O.

14

Gaseous waste control from coal-fired power stations

W.D. Halstead

14.1 INTRODUCTION

The impact of industrial society on the environment is the subject of increasing international interest and concern, a particular focus being the complex series of processes and phenomena generally grouped together under the generic title of 'Acid Rain'. This, in turn, is associated with the release of sulphur and nitrogen oxides into the atmosphere via the burning of fossil fuels. Power stations are major consumers of fossil fuels particularly, in the United Kingdom, coal, and methods for the control of their sulphur and nitrogen oxide emissions have been the subject of considerable and growing attention over the past three decades or so.

These developments have been paralleled by progressively increasing pressure to tighten legislative controls on emissions. This culminated within the European Economic Community with the development of a Framework for Action and then, in June 1988, with an agreement between the Environmental Ministers of the member states to a Directive on 'The limitation of emission of pollutants into the air from large combustion plants', (EEC, 1984, 1988). As a signatory to the Directive, the United Kingdom is committed to reducing its SO_2 and NO_x emissions and as large point sources of both, coal-fired power stations may be expected to make a major contribution to achieving the targets specified. This paper reviews the various technical options which are currently available or which might be suitable for future use in controlling emissions from these power stations. It also

The Treatment and Handling of Wastes
Edited by A.D. Bradshaw, Sir Richard Southwood and Sir Frederick Warner
Published in 1992 by Chapman & Hall, London, for The Royal Society
UK ISBN 0 412 39390 5, USA ISBN 0 442 31461 2

considers the implications of applying these technologies in terms of secondary effects within both the power station and the country as a whole.

14.2 LEGISLATIVE TARGETS FOR EMISSION CONTROL

The objectives of the Framework and Large Combustion Plant (LCP) Directive are to achieve overall reductions in national emissions of SO_2 and NO_x. As such, a degree of flexibility in national programmes is available in reaching a decision on the most cost-effective means of gaining the objective. This flexibility carries over to the contribution to be made by power stations in terms of factors such as the specific stations selected and the types of control to be fitted.

In 1987, prior to the agreement to the LCP Directive, the Central Electricity Generating Board (CEGB) had already announced its intention to retrofit flue gas desulphurization (FGD) equipment to 6000MW(e) of coal-fired boiler plant, which is the equivalent of three major power stations, to ensure that the existing level of SO_2 emissions would not increase in the future. At the same time, it was proposed that all new plants should be fitted with FGD. The consequences of the EEC Directive, which requires national emissions from all types of existing plant in the U.K. to be reduced in three stages relative to 1980, 20% by 1993, 40% by 1998 and 60% by 2003, could require an additional increment of FGD capacity, possibly as much as doubling the earlier voluntary retrofit programme to 12000MW(e). This is sensitive, however, to the degree to which alternative options, for example the construction of new plant, the use of low sulphur imported fuels or natural gas to replace existing fuels, can be applied in the next decade or so.

The requirements for NO_x are for reductions in U.K. emissions from large combustion plant of 15% by 1993 and 30% by 1998. Unlike SO_2, where power generation is responsible for some 72% of the total released in the U.K., for NO_x, the electricity industry produces only about 35% of the total emissions, road transport (45%) being the largest single source (Department of Environment, 1988). At the same time, power stations remain large point sources and since 1985 the CEGB has been carrying out a series of trials of commercial equipment to establish the extent to which NO_x formation, which occurs during the combustion process, can be suppressed by modifications to the burners. In 1987, the CEGB announced a ten-year programme to retrofit all its large coal-fired boilers with low NO_x burners. This involves the replacement of some 2000 burners in 44 boilers with a total generating capacity of 23000MW(e). This exercise is complicated by the variety of types of boiler currently in operation. Burners and firing patterns vary and each must be considered separately in identifying the optimum burner design.

14.3 THE CONTROL OF SO$_2$

14.3.1 Background

The main options for sulphur emission control are summarized in Fig. 14.1. Of these, the wet flue-gas scrubbing processes, which extract the sulphur dioxide from the flue gases just prior to their release from the power station, are by far the most widely employed, internationally. It is this type of technology which has been selected for the first U.K. installation on the 4000MW(e) power station at Drax in Yorkshire. Wet scrubbing processes will thus be considered first.

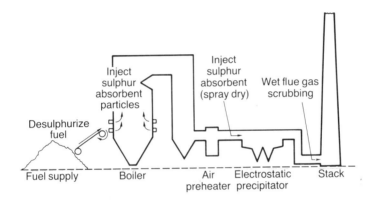

Figure 14.1 Main options for sulphur emission control.

14.3.2 The development of FGD by wet scrubbing

Wet scrubbing FGD units are basically large chemical plants in which an acid gas is reacted with an absorbent. A fundamentally simple concept, the first steps in the development of the technology were made within the United Kingdom over the period 1930–1960, principally at the Thames-side power stations at Battersea, Fulham and Bankside, where planning consents for station construction were qualified by a requirement to limit their sulphur emissions (Kyte *et al.*, 1981). Since then the pace has quickened on an international basis, so that over a hundred different types of process, albeit some quite closely related to each other, are now available, reflecting the ingenuity of the chemist and chemical engineer and the local cost and availability of particular reagents, (Behrens *et al.*, 1984). Degrees of development range from bench-top to 2000 MW(e) power station scale.

These processes can be classified in a variety of ways, but the most useful is probably one which relates to the way in which the principal absorbent is actually used. Taking this as the primary criterion and accepting that no split can be totally perfect, three classes of FGD emerge:

1. regenerative, in which the reagent is constantly regenerated with the sulphur recovered in a by-product as a concentrated stream;
2. once-through, in which the reagent is used only once, but emerges in an economically and/or environmentally useful form;
3. throw-away, where the main product has no economic or environmental value.

It is now useful to consider individual examples of each of these in somewhat greater detail.

14.3.3 Regenerative FGD

Any chemical cycle that allows sulphur dioxide to be efficiently absorbed in a process medium which can then, on saturation, be recovered by cycling through a different set of process conditions (e.g. by heating), may be used for this type of FGD. One process of this type which has emerged on the commercial scene is based on the sodium sulphite/sodium bisulphite cycle, as described in equation (1):

$$Na_2SO_3 + SO_2 + H_2O = 2NaHSO_3. \tag{1}$$

Known as the Wellman Lord process, a solution of sodium sulphite is used to scrub the flue gases; when saturated the liquor is transferred from the reactor to an evaporator-crystallizer, in which the reagent is recovered by steam stripping.

The basis of a second process is shown in equation (2):

$$Mg(OH)_2 + SO_2 = MgSO_3 + H_2O. \tag{2}$$

The reagent is an aqueous slurry of magnesium hydroxide and after reaction this is recovered via calcining of the reaction product.

For the operator regenerative systems have some attractions. In particular, since the reagent is constantly recycled, it minimizes the flow of materials into and out of the power station. In addition the sulphur can be recovered as a pure, concentrated stream of sulphur dioxide which is directly suitable as a feedstock for the production of sulphuric acid, sulphur or any other sulphur containing product for which a market can be identified.

14.3.4 Once-through FGD

This category includes the principal modern type of FGD technology which is based on the conversion of limestone to gypsum, the overall process being described by the equations:

$$CaCO_3 + SO_2 \rightarrow CaSO_3 + CO_2, \tag{3}$$
$$CaSO_3 + (O) + 2H_2O \rightarrow CaSO_42H_2O. \tag{4}$$

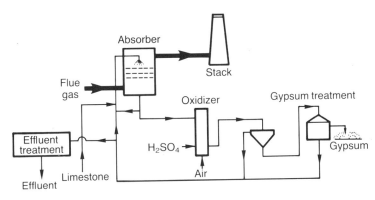

Figure 14.2 Schematic limestone/gypsum FGD Plant.

A schematic of this type of process is shown in Fig. 14.2. Lime is a technically viable alternative reagent but its costs, relative to limestone, make it unattractive in the U.K. context.

The final product, gypsum, has a number of industrial uses, for example, in wallboard manufacture. Materials input to and output from a power station are greater than for a regenerative process but from the point of view of considering new FGD installations one of its greatest attractions is that it is already being operated on a large scale in a number of countries.

Figure 14.3 Foundations of FGD installation at Drax power station taking shape.

Figure 14.4 Conceptual view of FGD installation at Drax power station.

There is a wealth of experience on which to draw to ensure efficient and reliable performance.

It is this type of process which has been selected for Drax (Figs 14.3 and 14.4), which, at 4000MW(e) (6 × 660MW(e) boilers), will be the largest FGD plant in the world when completed. It is planned that the plant will be commissioned in three phases each covering two boilers, with final completion by the middle of the decade.

An interesting hybrid of the regenerative and once-through systems is the Dual Alkali process in which the SO_2 is initially scrubbed with a liquor containing a sodium-based absorbent (e.g. NaOH or Na_2CO_3). This is then regenerated by the addition of lime to precipitate the less soluble calcium sulphate/sulphite. A particular advantage of this type of process is that it reduces the risk of scaling within the main SO_2 absorber. For present purposes, development of this type of process provides a useful demonstration of how a basically simple idea for an FGD process, the reaction of SO_2 with an absorbent, has been adapted to aid its successful application on the commercial scale.

14.3.5 Throw-away FGD

The beneficial utilization of as much FGD by-product as possible is a primary objective in the selection and operation of this type of plant within the United Kingdom. By comparison, many older installations in the U.S.A. which use lime or limestone absorbent, produce a difficult to de-water, thixotropic calcium sulphite–sulphate sludge (see equation (3)) as the final product. This is discharged directly to landfill, often with little attempt at further oxidation to gypsum (equation (4)). By mixing the sludge with pulverized fuel ash (pfa), excess lime, etc., improvements in the stability of the final product can be achieved but at best, the process demands substantial disposal areas.

Although sludging processes are not being considered for U.K. application one throw-away option which cannot, at this stage, be excluded exploits the natural alkalinity of seawater. The idea is similar to that utilized in the early Thames-side power stations. These are no longer operational but employed alkaline river water as the basic process medium, the water then being returned to the river. The main attractions of this type of process are the relatively low capital and operating costs and the minimization of by-product disposal problems. For power stations with access to suitable waters this process cannot be dismissed as a possible option for future installations. Proper control of discharge quality to ensure environmental compatibility will be a crucial preliminary requirement, however.

14.4 ALTERNATIVES TO WET FGD

14.4.1 Coal pretreatment

In addition to the wet flue gas scrubbing processes discussed above, there are a number of other options for the control of sulphur dioxide emissions, perhaps the most obvious being the use of fuels with lower sulphur content.

In pursuing this option the import of coals with lower suphur contents than those currently burned, which predominantly come from indigenous sources, is one possibility. An alternative would be to remove some of the sulphur in the coals currently used. These average 1.6wt% sulphur of which about half is present in the form of discrete iron pyrites (FeS_2) crystals. They have quite different properties from the organic coal matrix in which they are embedded, opening the way to separation by physical processing. British Coal have carried out a detailed examination of the prospects for extending their existing coal cleaning techniques to reduce pyritic sulphur in power station coal but found that at a cost competitive with wet FGD only 20 to 30kt sulphur could be recovered out of a total supply to the CEGB of the order of 1.2Mt.

The key factor in determining the economics was the cost of the losses of combustible material which occur during the cleaning process. To minimize these it is first necessary to grind the coal to liberate the pyrites (Fig. 14.5). In modern power stations the coal is finely ground before combustion and

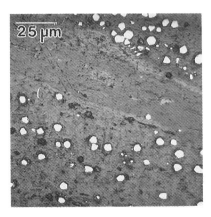

Figure 14.5 Iron pyrites (FeS$_2$) particle distribution in power station coal.

post-grinding processes exploiting differences in density, magnetic characteristics, conductivity, wettability, etc. between the organic matrix and pyrites in powdered coal have received widespread attention.

None of these possibilities can be regarded as having reached the commercially proven stage, however, and their attractiveness is diminished by their inability to recover, on average, more than half the sulphur content (e.g. the pyritic fraction) of the fuel. Chemical or biological treatment might achieve a greater degree of recovery but assessments indicate that such processes would be slow and involved, also producing a difficult to handle final product (wet, fine coal) and environmentally awkward by-products.

14.4.2 Combustion chamber control

Another possibility is to control sulphur emissions by the injection of an absorbent, e.g. limestone, directly into the combustion chamber. This reacts with the sulphur dioxide, the dry particulate product subsequently being collected downstream together with the pulverized fuel ash in the electrostatic precipitators. In practice, this type of process is only really suited to the new generation of burners designed to minimize nitrogen oxides formation by staging the combustion process. This staging reduces the top flame temperature, thus limiting the degree of deactivation (deadburning) to which the absorbent particles are susceptible when exposed to the high temperatures of conventional burners.

Even using low nitrogen oxides burners, however, detailed assessment of the LIMB (Limestone Injection into Multistage Burners) process shows it to be unattractive for U.K. conditions. Absorbent utilization is poor, as compared with wet scrubbing, and disposal outlets have to be found for the large volumes of partially sulphated reaction product, intimately mixed with the pfa, which are produced. A detailed assessment of this type of process led to the conclusion that it was unattractive for U.K. application (Burdett *et al.*, 1985).

14.4.3 Spray drying

While strictly a wet FGD process this is considerably different in concept to those discussed earlier, in that it involves the injection of a solution of absorbent (lime, sodium carbonate), in the form of a fine, aqueous spray, into the flue gases up-stream of the electrostatic precipitators. This finely divided dispersion has a high reactivity which boosts both absorbent utilization and sulphur take-up efficiency relative to the furnace-based technique, though the final product remains a mixture of partially reacted reagent. Depending on plant design, it may also be intimately mixed in which the pfa. This then has to be disposed of in an environmentally acceptable manner.

14.4.4 Advanced combustion technologies

In addition to the SO_2 emission control technologies currently available for conventional boiler plant, considerable effort is being made to identify new combustion techniques which combine improved thermodynamic conversion efficiencies with a substantial emission control capability. The improved efficiency stems from their combined cycle mode of operation based on gas as well as conventional steam turbines.

Within the U.K., the CEGB and British Coal have recently completed a £28M joint research programme on 'pressurized fluidized bed combustion' which exploits this concept. Direct sulphur dioxide control is achieved by mixing an absorbent (e.g. limestone) directly into the fluidized bed in which the coal is burned (Fig. 14.6). Fluidization is achieved via the combustion air which enters the base of the bed. The basic concept can be applied in a number of different ways, to meet the specific needs of particular operators in industrial as well as power generation markets.

A second option receiving detailed attention within the U.K. involves gasification of the coal, the fuel gas product then being cleaned (e.g. de-sulphurized) before, again, being burned in a combined cycle plant (Fig. 14.7).

In addition to the experimental programme two design studies have been published by the Department of Energy (Department of Energy, 1988). In one, the CEGB collaborated with British Coal and the Department of Energy in an engineering assessment of combined cycles based on fluidized bed combustors. The second with British Coal, British Gas and the Department of Energy covered the application of the British Gas/Lurgi

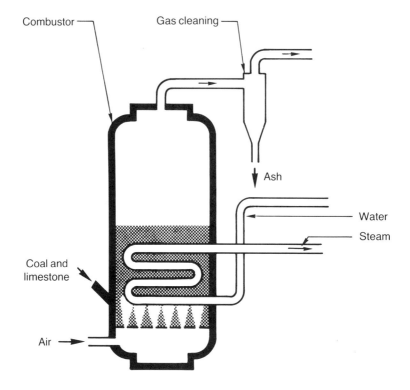

Figure 14.6 Pressurized fluidized bed combustion system.

slagging gasifier to large scale combined cycles. It assumed that the necessary gas turbines will become commercially available in the 1990s.

The studies indicate that under U.K. conditions and assuming state industry rates of return for the designs studied:

1. at unit sizes below 400–600 MW(e), depending on the economic assumptions, pressurized fluidized bed combustion may produce electricity at a lower overall cost than pulverized fuel combustion linked to FGD. However, this would still be at higher cost than larger (660 MW(e) or more) pulverized fuel plus FGD units unless the waste heat could be sold at a sufficiently high price (e.g. via combined heat and power);
2. an integrated gas fired combined cycle plant design based on the British Gas/Lurgi slagging gasifier shows promise of being able to compete with conventional plant with FGD at sizes of 660–900 MW(e), assuming continuing improvements in gas turbine efficiency and a market for the sulphur by-product.

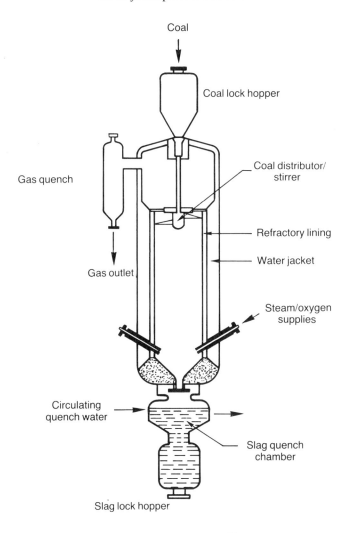

Figure 14.7 Slagging gasifier.

14.5 DEVELOPMENT OF THE U.K. FGD INSTALLATION PROGRAMME

Development of an installation programme of the size now being considered for the United Kingdom is an extremely complex and demanding exercise. A diversity of interlinked issues must be taken into account to establish the optimum course of action, in terms both of types of FGD plant to be installed and of the most suitable power stations to receive them. Some of the issues are related to the general conditions pertaining in the U.K., e.g. the size and age distributions of the stations and the characteristics of the coals they burn (for example, their sulphur and chlorine contents), others pertain to the conditions existing at particular power stations (e.g. ease of access to road/rail transport, station design, local availability of raw materials such as limestone, local outlets for by-products (e.g. gypsum)).

It was as a result of such detailed analysis that the decision was taken that Drax, which has the newest coal-fired generating sets in the country, should be the first station to be retrofitted with FGD, the well proven limestone/gypsum process being selected. It is planned that the by-product gypsum should be used principally to manufacture building products, especially wallboard, an indication of the logistics involved being given in Table 14.1.

Table 14.1 Estimated materials inputs and outputs for an FGD plant retrofitted to Drax units 4–6 (1.7wt% S coal, 82% load factor)

		Quantity (tonnes per annum)	
Imports		Exports	
Limestone	240000	Gypsum (limestone/gypsum)	430000
		or sulphuric acid	220000[a]
		or sulphur (regenerative)	75000[a]
		Impure gysum (from waste-water treatment plant)	10000–50000

[a] Comparative Regenerative FGD Output.

As the number of FGD-fitted power stations increases, the opportunities for disposing of particular bulk by-products directly to industry will diminish, and different criteria will begin to impose themselves in selecting the types of process for further installation. For example, the gypsum output from one 2000MW(e) power station would be equivalent to 15–20% of the national market and questions need to be asked about the degree to which FGD gypsum could, as production builds up, be expected to displace gypsum from other sources. The higher the proportion the greater would be the receiver companies' dependence on FGD output and the greater, in turn, would be the producers' commitments in terms of continuity and quality of supply.

This could open the way for the application of other types of FGD on later retrofits, for example, regenerative units producing sulphur, concentrated SO_2 or H_2SO_4 as the final product. The output from a 2000MW(e) sulphur or H_2SO_4 producing plant is also set out in Table 14.1. This level of acid production equates to 10–12% of the U.K. market, but the market is characterized by the close integration of supply and utilization. This means, in turn, that opportunities for the marketing of FGD acid would be restricted with concomitant implications for the application of this type of FGD process.

As the FGD installation programme progresses, it must therefore be anticipated that a stage will be reached at which further supplies of by-product cannot be absorbed by industry. To meet this challenge, the CEGB, in collaboration with various academic and other specialist groups (e.g. Forestry Commission: Agricultural Advisory and Development Service of the Ministry of Agriculture, Fisheries and Food) is evaluating a range of environmentally beneficial alternative disposal options for gypsum. These include the recovery of low grade saline and other soils and, more speculatively, and in combination with pulverized fuel ash and cement, via the manufacture of building blocks for the construction of artificial reefs. Studies targetted at establishing the landfill options for gypsum disposal are also underway. These involve determining its mechanical and hydrological characteristics, both alone and when mixed with ash, and the development of methods for recovering landfill sites by revegetation for amenity or other uses (Fig. 14.8). It is a key objective to ensure that the FGD installation programme is not impeded by problems arising from by-product disposal (Halstead, 1988).

The CEGB is also researching, at both the laboratory and pilot plant scale (Fig. 14.9), a range of issues relating to the operation of the FGD plant itself.

Figure 14.8 FGD gypsum experimental landfill site at Drax power station.

Figure 14.9 CEGB FGD experimental facility.

While plant are commercially available up to the largest sizes of interest from several vendors, the design of each unit must be individually tailored for optimum performance under the specific range of conditions which apply at the installation site. Prominent among the current topics being studied are:

1. relative reactivities and purities of the numerous candidate limestone feedstocks for U.K. limestone/gypsum FGD application;
2. means of controlling the internationally high levels of HC1 in U.K. boiler flue gases and their effects on FGD performance;
3. performance of materials of construction (steels, protective coatings) under U.K. FGD conditions.

14.6 CONTROL OF NO$_x$

14.6.1 Background

The concentration of nitrogen oxides in the flue gases of coal-fired power stations varies considerably with furnace design, with nitric oxide, NO, generally accounting for over 90% of the total. Unlike sulphur dioxide where the only source of sulphur is the fuel, the parent nitrogen has a second source in the combustion air, the combustion temperatures produced by modern power station burners (>1700K) being sufficiently high for a degree of direct oxidation to take place. NO$_x$ produced from atmospheric nitrogen is generally referred to as 'thermal' NO$_x$.

'Fuel' NO$_x$ is formed in parallel with 'thermal' NO$_x$ by the oxidation of nitrogen chemically bonded within the coal, in which it is usually present in the 1.0–1.3wt% concentration range. Not all the fuel nitrogen is converted to NO$_x$, however. Elementary nitrogen is a second possible product, the degree to which one or other is favoured depending, in part, on the precise conditions pertaining within the combustion system. Essentially there are two possibilities, shown in Fig. 14.10. In modern power stations, coal is generally burned in the form of a fine powder to promote efficient, high intensity combustion. The first stage of this process entails the heating up of the particles as they enter the combustion zone, with release of the volatile components including much of the nitrogen containing material. It is this

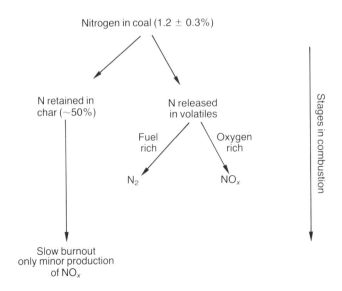

Figure 14.10 Formation of NO$_x$ during combustion.

nitrogen which is then oxidized to produce NO_x. The devolatilized residual char, containing further nitrogen, also burns away but more slowly and at lower temperatures, conditions that are less favourable to the formation of NO_x. Of the NO_x emitted by power stations, 'fuel' NO_x predominates, an indicative figure being 70% of the total though this is sensitive to a variety of factors including the exact level of nitrogen in the fuel, the flame temperature, residence time and oxygen level in the reaction system.

14.6.2 Control of NO_x by combustion modification

The extent to which NO_x is produced in the flue gases of modern power stations is influenced by the precise conditions occurring within the combustion zone of the burners being used. It is reasonable to anticipate, therefore, that opportunities will exist for developing improved burner types which enable reductions in NO_x production rates, relative to standard designs for which NO_x control was not a special concern. It is on the development of improved, 'low NO_x' burners that attention has, in fact, concentrated for the control of NO_x emissions.

Standard power station burners are designed to achieve rapid, efficient

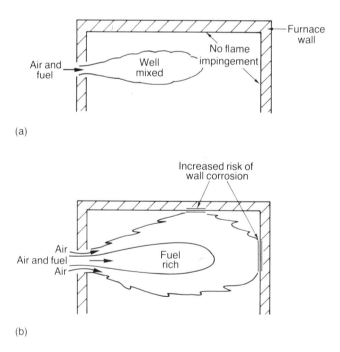

Figure 14.11 Multistage combusion in low NO_x burner: (*a*) high intensity standard burner; (*b*) low NO_x burner.

combustion via good mixing of the fuel particles and combustion air. Low NO$_x$ burners seek to achieve the same degree of combustion but in a staged manner as shown in Fig. 14.11. In effect the combustion air is divided, the core of the flame being made from the fuel mixed with a proportion of the combustion air. Partial oxidation of the fuel takes place within this zone but with a sub-stoichiometric level of air in the vicinity. Combustion is then completed by further reaction as the partially oxidized combustion products come into contact with an outer envelope made up of the remainder of the combustion air, the 'secondary' air, injected around the central zone. This attenuation, in effect, reduces the rate of the combustion process limiting the maximum temperature and suppressing NO$_x$ formation. The challenge for the combustion engineer is to ensure that such burners achieve the same degree of fuel conversion as standard burners (e.g. fuel is expensive and any loss of conversion efficiency will be a direct charge on the cost of the final product, electricity) and that the burners can be successfully accommodated within the power station boilers for which they are intended.

14.6.3 Programme for low NO$_x$ burner development

To achieve national NO$_x$ emission targets, low NO$_x$ burners are being retrofitted to all the CEGB's largest boilers (e.g. $\geqslant$500MW(e)). Several different types of large boiler are in operation, however, and it is necessary to develop individual burner designs which will be compatible with each. As discussed, significant reductions in NO$_x$ production rates and effectively achieving 'burn-out' of the fuel are common requirements to all new burner types, but in tailoring them to individual boilers a further factor needs to be taken into account. The boilers themselves were originally designed and sized to accommodate standard burners, but attenuation of the combustion process heightens the risk of exposing adjacent walls of the furnaces to flame impingement, the deposition of partially combusted fuel particles, etc. Under such conditions the walls, which are generally lined with steel tubes through which water is passed in the early stages of the steam raising cycle, will be rapidly corroded with severe and expensive consequences in terms of boiler failure and loss of availability. Whatever the merits of particular burner designs in controlling NO$_x$), the total retrofit system, when brought into operation, must be so designed as to ensure that adequate protection against furnace wall corrosion is assured.

There are, in fact, three basic types of large power station boiler operated by the CEGB, which may be classified according to burner arrangement (see Fig. 14.12) as Tangential, Front Wall or Opposed Wall fired. The Front Wall group may be divided into two sub-groups, differentiated by the size of the individual burners used. The CEGB has a programme of burner development for each individual boiler type, attention initially concentrating on tangentially fired boilers with a 500MW(e) boiler at Fiddlers Ferry Power

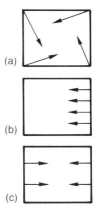

(a)

(b)

(c)

Figure 14.12 Firing patterns on large, modern power station boiler plant: (*a*) corner fired, e.g. Fiddlers Ferry; (*b*) front-wall fired, e.g. Eggborough; (*c*) opposed fired, e.g. Drax.

Station acting as the test-bed. Tangentially fired boilers are probably the most straightforward to retrofit in that their firing pattern results in the combustion process being concentrated in a fireball at the centre of the furnace, away from the walls which can also be protected by provision of an air curtain. The Fiddlers Ferry programme has been very successful with NO_x reductions of the order of 35% being achieved, relative to the standard burner operational range of 470–550 ppm (at 3% excess O_2).

Development programmes for low NO_x burners on Front Wall (Large Burner) and Opposed Wall Fired boilers are now well underway at Eggborough and Drax Power Stations, with work on the final family, Front Wall Fired (Small Burner) scheduled to start soon.

14.6.4 Post-combustion NO_x control

While modifications to burner technology can be looked for to give tangible reductions in the amounts of NO_x produced relative to standard burners, there are limits to the degree of reduction which can be achieved from this kind of development. In addition to improving burner technology attention is thus also being given, on a worldwide basis, to possible downstream control methods. The basic concept on which most of these rely is that of reacting the NO_x present in the flue gas with a reducing agent to produce nitrogen as the final product. Practical interpretation of this simple, general concept has been via a variety of different routes, as outlined in Fig. 14.13, however.

At one extreme reduction can take place within the furnace itself by a

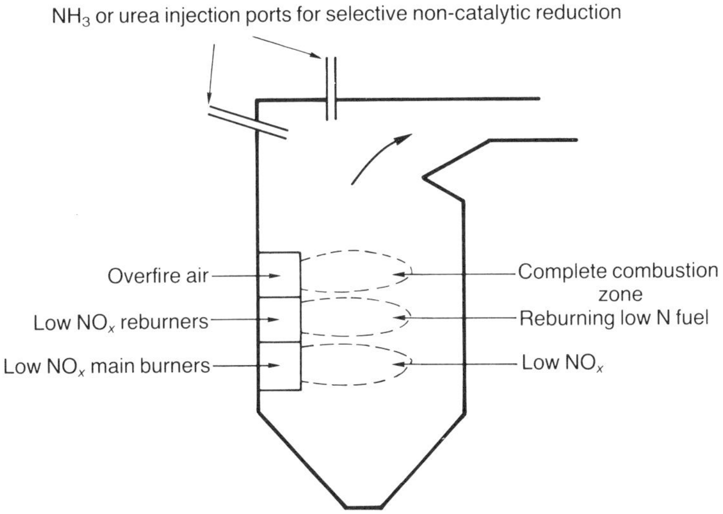

NH$_3$ or urea injection ports for selective non-catalytic reduction

Figure 14.13 In-furnace and SNCR NO$_x$ reduction.

method similar in some ways to that employed within low NO$_x$ burners. With In Furnace NO$_x$ Reduction (INFR) the principal, low NO$_x$ burners occupy the lower ranks of burner ports within the furnace, the upper ranks accommodating in ascending order burners operating fuel rich, preferably with a low nitrogen fuel (such as natural gas) and, at the top, overfire air supplies. This sequence promotes reduction of the NO$_x$ within the flue gases produced by the main burners, via mixing with the oxygen deficient (e.g. reducing) flue gases coming from the upper burners, oxidation of the total fuel input then being completed by the overfire air.

A second option involves injecting a specific reducing agent into the flue gas, for example ammonia or urea, which then, again, reacts with and reduces the NO$_x$. The effectiveness of this type of process is sensitive to temperature. If it is too high at the point of injection the reducing agent is itself oxidized to further NO$_x$ by free oxygen in the flue gases (economic operation of boiler plant dictates an overall excess of air to ensure efficient burnout of the fuel). Conversely, if the temperature at the injection point is too low, then insufficient precursor radicals (NH$_2$:OH) are formed via dissociation of the initial additive, for the NO$_x$ reduction process to be fully successful, (Miller *et al.*, 1981).

Known as Selective Non-Catalytic Reduction, (SNCR) the flue gas temperature window for this type of process lies between 1400 and 1100K, which places it within the upper reaches of the furnace. This is a crowded area which accommodates a complex array of superheater and reheater tubes and is thus not conducive to the incorporation of injector banks. Care also has to be taken to minimize the degree to which unreacted additive escapes the reaction zone, since its subsequent reaction with other components of the flue gas, e.g. SO_2, SO_3, HCl could lead to deposition of reaction products and hence fouling of downstream plant surfaces (e.g. of air heaters) which would influence the overall performance of the plant. As with INFR, SNCR will require a substantial development programme before it can be regarded as a commercially viable option for large CEGB boiler plant.

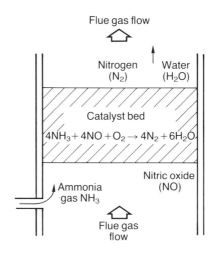

Figure 14.14 Dry flue gas De NO_x.

A further alternative is selective catalytic reduction (SCR) of NO_x (Fig. 14.14). This operates at flue gas temperatures in the range 600–700K and again involves reducing the NO_x by using a flue gas additive, e.g. NH_3, although this time in the presence of a catalyst,

$$4NO + 4NH_3 + O_2 \rightarrow 4N_2 + 6H_2O \tag{5}$$

At the operational temperatures, the flue gases are still upstream of the electrostatic precipitators which remove from the gas stream the heavy loading of fine coal ash particles which become entrained during the combustion process. Fouling of the catalyst surfaces by deposited ash could seriously inhibit the effectiveness of SCR and one of the key features in its development has been the evolution of catalyst support structures which are aerodynamically resistant to the build up of ash deposits. A number of different types have been evolved ranging from honeycombs to packed layers of corrugated plates.

The catalysts themselves are generally proprietary mixtures of transition metal compounds which could be poisoned by other components of the flue gases. It is thus necessary, before considering for use in any particular plant, to show that acceptable catalyst life times can be achieved under the conditions pertaining. This is particularly true of the U.K. situation where, for example, power station coals have exceptionally high levels of chlorine (average 0.25wt%, range 0.1–0.7wt%) relative to those used elsewhere in the world (e.g. Japan, West Germany). Up to 90% NO_x reduction is achievable under favourable conditions, but rigorous testing in CEGB flue gases will be needed before this control option could be confirmed for U.K. application.

Another problem of this type of process is that it requires the insertion of a large volume of catalyst (up to 2000m^3 for a 2000MW(e) (power station), into the power station structure. Making space for this and ensuring that it is maintained within the requisite temperature range would be a major problem on an operational power station, as would coping with any slippages of unreacted additive which could lead to the fouling of downstream surfaces.

14.6.5 Combined SO_x/NO_x control options

Control of SO_2 and NO_x at the commercial application level has, in large measure, been regarded as a two stage process with separate technologies (e.g. burner modifications, flue gas washing) targeted at the individual species. Attention is, however, being given to the prospect for combining the SO_x/NO_x control processes and a wide variety of possibilities have been and are being researched.

In some cases, these are 'dry' methods involving the adsorption of SO_2 onto a high surface area material which can also be used to catalyse the reduction of NO_x. One such system employs activated carbon which adsorbs the SO_2 during which it is oxidized, e.g.

$$SO_2 + \tfrac{1}{2}\,O_2 + H_2O \rightarrow H_2SO_4 \text{ (adsorbed).} \tag{6}$$

The carbon can subsequently be recovered by heating, during which partical oxidation takes place, e.g.

$$2H_2SO_4 \text{ (adsorbed)} + C \rightarrow 2SO_2 + CO_2 + 2H_2O. \tag{7}$$

The carbon, plus a catalyst, is also suitable to promote the reduction of NO_x using an additive, e.g. NH_3.

The NOXSO process, by contrast, uses a solid sorbent produced by depositing sodium carbonate on an alumina substrate. Reaction takes place to produce $NaAlO_2$, which is then able to absorb SO_2, the overall process being described by equations (8) and (9):

$$2NaAlO_2 + SO_2 \rightarrow Na_2SO_3.Al_2O_3, \tag{8}$$

$$2NaAlO_2 + SO_3 \rightarrow Na_2SO_4.Al_2O_3. \tag{9}$$

The $NaAlO_2$ also absorbs NO from the gas stream. Regeneration of the absorbent is then achieved by heating to about 800K in the presence of a reducing gas (e.g. H_2 or CO) to dissociate the more stable reaction products (Keeth, Krajewski and Ireland, 1986).

Another group of processes which are considerably different in concept, involve the gas phase oxidation of SO_2 and NO to SO_3 and NO_2 by, for example, electron bombardment. The gas stream is then neutralized with an alkali, e.g. NH_3, either in the gas phase or via a scrubber, to produce an $(NH_4)_2SO_4$ based final product for which, after collection from the gas stream, commercial markets might be found. A key issue with this type of process is the persistent plume which can be produced from the plant if the collection process is not of very high efficiency.

It remains to be seen whether any of these new processes will, in time, supercede the proven technologies now commercially available. What is clear, however, is that unequivocal demonstration of performance reliability will be necessary before utilities operating in a competitive commercial environment will be able to justify their application in preference to proven existing technologies.

14.7 CONCLUSIONS

The development of modern society is based, in large measure, on the availability and reliability of electricity supplies. These are produced, in the main, via the combustion of fossil fuels, in particular in the United Kingdom from coal. At the same time, concern for the protection of the environment, with its concomitant tightening of legislative controls, is a particular feature of that society. This has created major new challenges for electricity utilities worldwide, who must ensure that the impact of their operations on the environment is kept within acceptable limits. In effect it has acted as a spur to the conception and development of a wide range of control processes. These reflect, in particular but by no means exclusively, the ingenuity of chemists and chemical engineers in developing methods for controlling and in some cases exploiting for beneficial use, the more environmentally sensitive by-products of electricity production, most specifically SO_2 and NO_x.

These two problem by-products provide an illustration of the alternative approach being considered. With NO_x the aim is to convert it into a totally innocuous material (N_2) which can then be released without worry. With SO_2 this is not possible, so instead, it is converted into materials that can be deposited safely. The materials produced can be of commercial value so that in the end the ideal solution is found, where there is nothing to be deposited at all.

The developments involved range from modifying the combustion system by degrees extending from relatively straightforward adaptations of existing burner systems to total rethinks of combustion technology, (e.g. fluidized beds and gasification/combined cycles), through to the large-scale processing of flue gases before discharge. Applications of these processes, in themselves, often bring with them secondary environmental effects related, for example, to the supply of suitable feedstocks and the disposal of by-products. Such issues must also be satisfactorily resolved before the primary process can be seriously considered for use.

As a result of this activity a range of emission control technologies have now been developed to and proved on the commercial scale, and are being applied in an increasing number of countries including the United Kingdom, notwithstanding the costs involved (e.g. the 4000MW(e) FGD installation at Drax Power Station will cost £600M to £700M to construct) and the greater technical complexity which they bring to the overall electricity supply process. At the same time, the challenge continues to improve on existing control technologies by reducing costs, increasing efficiency and curtailing secondary environmental impacts. These are goals which will remain for as long as coal, or any other fossil fuel, is used in the production of power.

ACKNOWLEDGEMENTS

The author wishes to thank Mr M.E. Brown and Mr M.S.A. Skinner for their valuable comments during the preparation of this paper, which is published by permission of the National Power Division of the Central Electricity Generating Board.

REFERENCES

Behrens G.P., Jones G.D., Meserole N.P. and Seames W.S. (1984) *The evaluation and status of flue gas desulphurisation systems.* EPRI Report CS 3322.
Burdett N.A., Cooper J.R.P., Dearnley S., Kyte W.S. and Tunnicliffe M.F., (1985) The application of direct limestone injection to U.K. power stations *J. Inst. Energy* **58**, 435.
Department of Energy (1988) *Prospects for the use of advanced coal based power generation plant in the United Kingdom.* Energy Paper no. 56. London: H.M.S.O.
Department of the Environment (1988) *Digest of environmental protection and water statistics*, no. 11, Government Statistical Service. London: H.M.S.O.
EEC Framework Directive 84/360/EEC.
EEC Large Combustion Plant Directive 88/609/EEC.
Halstead W.D. (1988) Flue gas desulphurisation within the CEGB; objectives and consequences. In Proceedings of symposium *New Frontiers for By-Product Gypsum.* Ontario: Ortech International.
Keeth R.J., Krajewski M.J. and Ireland P.A. (1986) Economic evaluation of FGD systems. *EPRI Report CS* 3342, **5**.
Kyte W.S., Bettelheim J. and Littler A. (1981) Fifty years of experience of flue gas desulphurisation at power stations in the United Kingdom *Chem. Engineer* **369**, 275.

Part Six

Conclusions

Chairman's introduction

T.R.E. Southwood

What are we to conclude about the technology for the handling and treatment of wastes in the third millennium? Although, as we have heard (McInerney, Chapter 12) in relation to radioactive wastes, calculations have been made for storage periods of tens of thousands, even tens of millions of years, such time scales are well beyond the range of reasonable forecasts. There is one respect in which it is valuable to think against such a long timescale: it reminds us that our actions now may have consequences well into the future. As has been said several times in the meeting 'we have borrowed the world from our descendents'. The extent to which waste will have long-term effects will depend on its properties, more precisely on the natural residence times of the substances involved. This is well illustrated by the distribution of the caesium and plutonium discharged from Sellafield (Woodhead, Chapter 9); a relatively brief period in the mid- to late-1970s when the quantity of radionuclides discharged was much greater than the current level, has given a legacy of plutonium that will remain for centuries. However, if there are any problems due to Cs-137 (with a short half-life) these will impinge on us now, on the generation, if not the actual people, responsible. I will return to the subject of residence times below, but in respect of most waste disposal issues we can do little more than peep into the next century, the whole millennium is rather beyond us!

The first step in forecasting must be to consider the general background against which these technologies will be developing. We can assume, perhaps optimistically, a continued upward movement of GNP worldwide, and more pessimistically that world population will also grow. As we have been reminded in the discussions these trends will have considerable repercussions in countries such as China and India with the growth of their energy demands and of their other needs, for example for refrigeration and thus for CFCs. Frank Fraser Darling stated in his 1969 Reith Lectures 'most pollution comes from getting rid of wastes at the least possible costs', although from more recent experience (Bringer, Chapter 4), we should rephrase this to 'most pollution comes from getting rid of wastes at what appears to be the least possible cost'. So left unconstrained, pollution from wastes could clearly become much worse as we enter the next millennium. Fortunately, we can put against that trend the growing public understanding and interest in environmental matters, the general 'greening' of politics. This seems likely to provide constraints or, as I think we should see it, the pressures necessary to ensure that the best technological solutions are applied to the problem of waste. This means that in the manufacturing industry, the reduction of waste will have to be seen as an integral part of the production process. Pullen (Chapter 13) has pointed that the public will need to be kept informed with reliable information, 'handwaving excuses will no longer do'; once public confidence has been lost it is difficult to regain.

In the 11th Report of the Royal Commission on Environmental Pollution (RCEP) 'Managing waste: the duty of care' (1985) a strategy for waste management was developed. It included these stages:

1. Reducing the amount of waste, e.g. low-waste technologies and longer product life (Bringer, Chapter 4). One may hope and expect that the greater precision possible in manufacturing process through electronic control will be important in this respect (Suckling, chairman's introduction).
2. Maintaining a secure waste stream. This is essential if the next step is to be fully effective and a proper price paid for the appropriate disposal technology.
3. Disposal, which itself has two components: recycling and containment.

When we talk of 'recycling', we must remember that most waste is, and has been for millennia, re-cycled through the natural biogeochemical cycles. If a substance is recycled is it to be regarded as a pollutant? Kramer (Chapter 3) has expressed the view that it would be necessary to enact legislation containing lists of substances that are to be regarded as dangerous, for example dioxin which he maintained should be regarded as harmful until proved otherwise. Furthermore he has criticized the definition of 'harm' that was related to the impact of the substance on living organisms (Warner, chairman's introduction) citing the damage caused by pollution to the ozone layer and to ground water. I find it impossible to recognize the concept of 'harm' to these physico-chemical systems by a modification of their composition. The 'harm' arises from the consequent impact on living organisms, in these examples, of increased UV-B radiation or of toxic substances in drinking water, respectively. If the view is to be taken in drafting future EC regulations that the composition of physical systems is to be held constant then there will be much unnecessary expenditure. Such an outlook totally disregards the many natural processes, recycling and detoxifying, that serve to maintain the stability of our environment; the requirement is to avoid overloading them because then changes occur that are detrimental to living organisms (Chester, chairman's introduction).

Thus in assessing the properties of a substance before its inclusion in any list or in relation to the appropriate expenditure to control its level, it is important to consider features other than simply toxicity. Of course, toxicity is important, but residence time in the environment is another, often more significant property. As Bradshaw (Chapter 1) has reminded us, in Nature, heavy metals are mostly well locked into the deep layers of the Earth; mining brings them up to the surface and spreads them around. They will not easily be removed from the surface as their residence time is long and they are generally toxic in all their forms (though particularly so as organometallics) (Southwood, 1986). In contrast, natural organic compounds consisting mainly of carbon, oxygen and hydrogen may be broken down or changed by microbiol agents or other natural processes. Carbon monoxide is very toxic, but it has an extremely short environmental residence time.

Thus for degradable materials, contrary to views expressed several times at this meeting, dispersal may be an acceptable option, provided that the material is distributed in such a way as to avoid overloading the system and permit recycling (Hall, Chapter 5). One must confirm that this is the case. The method has been discredited because it has simply been assumed that the natural system would cope. Just as the new technologies of process engineering involving electronics should reduce the size of the waste stream, so we may look, as Bull (Chapter 10) showed, to biotechnology in its broadest sense, to contribute to an increased ability to strengthen or augment the natural processes by the use of particular microbial agents. That is to recycle, into the natural biogeochemical cycles. both natural and xenobiotic compounds. Anderson (Chapter 11) and Pullen (Chapter 13) point out how an efficient waste disposal plant should use both microbial and chemical methods for breakdown and recycling as well as containment. Industry will also develop more of its own cycles with wastes being used for other purposes (Halstead, Chapter 14), so alongside the natural cycles, we will have 'industrial cycles' (Frosch, Chapter 16); in this way the size of the waste stream will be reduced. For those wastes, containing substances that are so stable that they cannot be converted into more useful or less harmful materials, there is general agreement that they should be confined.

For all wastes these different options, recycling or containment, must be considered within the framework of the Best Practical Environmental Option. To do this objectively we need a great deal of information (Roberts, Chapter 15).

My own prognosis is reasonably optimistic. Despite increasing GNP and population, technology will come up with the answers needed, indeed demanded, by public opinion. There will be four requirements for a rational development:

1. A blend of sciences will be necessary to provide the basic understanding: engineers and biological and physical scientists need to work together (Rae, Chapter 8; Smith, Chapter 7). Solutions arrived at entirely within the framework of one discipline are likely to be sub-optimal. There has been a growing trend towards the introduction of interdisciplinary options in many higher education courses and the training of Environmental Health Offices is strongly interdisciplinary. But perhaps even more attention should be given in the training of single discipline professions to ensure that the practitioneers appreciate the potential contribution of other subjects and the limitations of their own (Suckling, chairman's introduction).

2. Professionals working in this field will have to be scrupulous in ensuring the scientific integrity of their statements and will have to be willing to share their doubts with the public (RCEP, 1984). With environmental matters there are often genuine uncertainties and one finds experts performing more as advocates than dispassionate scholars. On one hand I am concerned at the pressures that may be placed on scientists employed by commercial organizations; they may be required to avoid certain research or to make statements that are economical with the truth. Former students have too often complained to me of these pressures for me to doubt that they occur. On the other hand there are those independent spirits who are so anxious for the future that they may exaggerate the level or certainity of a risk, thereby causing considerable and potentially unnecessary anxiety. If scientists fail to maintain their integrity and so justify public reliance in their pronouncements then we can expect the greater involvement of the law in settling such issues, as in the U.S.A. (Jasanoff, 1986). Alternatively, virtually all options become blocked because of public disapproval as Backman (Chapter 2) has reported is occurring in Sweden in respect of waste treatment procedures. The development of new techniques for radioactive waste disposal in Britain is another example of policy paralysis due to public pressure (McInerney, Chapter 12). Unfortunately no amount of public concern will make these problems disappear! It is certainly not in the long term interests of industry that their scientists are distrusted, because sharing information with the local community is one of the key approaches to overcoming the NIMBY syndrome (RCEP, 1985). Secrecy fuels fears and once public confidence has been lost, its recovery is both lengthy and expensive.

3. More resources will have to be allocated to waste management (Turner, Chapter 6). Ultimately, even with the polluter pays principle being applied, the increased costs have to be borne by society as a whole, although as Bringer (Chapter 4) points out sometimes changes in procedures can bring down total costs. Certainly the costs of retrospective measures are likely to be far greater than those of proper disposal at the

time of production. Adequate inspection and enforcement are important factors in public confidence and therefore sufficient public funds will need to be allocated for these purposes. Public opinion surveys tend to support the commitment of more resources to environmental protection.

4. It will be necessary to apply longer time horizons than is common at present. First, the planning of new procedures needs to be associated with the development of new plant (Halstead, Chapter 14). In this way the costs of new environmental measures may be minimized, the Best Environmental Timetable (RCEP, 1985). The extent to which energy supply industries may be subject to short term financial pressures can be a particularly serious obstacle to the most cost effective pollution control programme for this major waste producing sector of the world economy. Secondly, waste will have to be cared for over long periods of time. This extended duty of care will apply to domestic and normal industrial waste in terms of decades (Anderson, Chapter 11; Pullen, Chapter 13) and to some radioactive wastes over periods that extend well beyond the next millennium.

REFERENCES

Jasanoff, S. (1986) *Risk management and political culture*. New York: Russell Sage Foundation.

Royal Commission on Environmental Pollution (1984) *Tackling pollution – experience and prospects*. 10th Report, Cmnd. 9149. London: H.M.S.O.

Royal Commission on Environmental Pollution (1985) *Managing waste: the duty of care*. 11th Report, Cmnd. 9675. London: H.M.S.O.

Southwood, T.R.E. (1986) Lead in the environment. In *The assessment of environmental problems* (ed. G.R. Conway, R. Macrory and J. Pretty), pp. 11–21. London: Imperial College.

15

Some outstanding problems

L.E.J. Roberts

15.1 INTRODUCTION

The principles on which sound waste management should be based have been much debated and clarified in recent years. With the growth of interest in environmental control generally, the rise in the 'green' vote, the recent international conference on pollution of the North Sea, and the wider interpretation of the costs of pollution, there will be increasing pressure to put these principles into effect. Some of the general problems that will emerge are enumerated in this paper, which deals particularly with the disposal of solid and toxic wastes. Some of these have to be seen in an international dimension. The control of atmospheric pollution, particularly if this extends to the limitation of carbon dioxide emissions, will entail even broader considerations of international policy and economics since the question of the means of energy supply will be involved.

The traditional, time-honoured principle of managing potential sources of pollution or nuisance in the U.K. is that they should be reduced by the 'best practicable means', BPM. The definition of what is practicable brings in questions of economics as well as of technology. The newer, more explicit formulation is BATNEEC, 'best available technology not entailing excessive costs'. The Health and Safety Executive, in considering regulations affecting risks from potentially hazardous installations, requires risks to be reduced to values 'as low as reasonably practicable' (ALARP); where 'reasonable' is again defined by the rule that risks should be reduced until the costs of further reduction are disproportionate to the additional benefit achieved. Both BATNEEC and ALARP put the onus on management to reduce risks

The Treatment and Handling of Wastes
Edited by A.D. Bradshaw, Sir Richard Southwood and Sir Frederick Warner
Published in 1992 by Chapman & Hall, London, for The Royal Society
UK ISBN 0 412 39390 5, USA ISBN 0 442 31461 2

if the means are available, even if the risks are low, until the cost of going further is excessive. But they are usually applied to the effects on a single medium, air or water. The newer and more revolutionary concept is the concept of the 'best practicable environmental option', BPEO, which recognizes the cross-boundary nature of pollution problems, as does the formation of a combined Pollution Inspectorate. Integrated pollution control seems to be here to stay (O'Riordan, 1989).

The adoption of these modern principles of waste management is being forced on our societies by the pressures of a diversifying inventory of wastes of all types, a growing concern for the environment, and a shrinking area of available land; 95% of total waste arisings are consigned to landfill or deposited on land in the U.K. (RCEP, 1985). The most obvious and often most economically efficient way of minimizing the toxic waste problem is to phase out processes generating toxic material in favour of others, and to recycle material wherever possible, as is argued in several papers to this Symposium. Such practices will reduce some waste streams and particularly industrial wastes, but will not eliminate waste disposal problems. The objective of a BPEO policy is to evaluate several options of waste management so as to choose the procedure which minimizes environmental pollution as a whole, taking economic considerations into due account, and not simply to minimize emissions to one environmental option. The Royal Commission (1988) summarized the steps concerned in selecting a BPEO, of which the most taxing are the following: 'Step 3: evaluate the options. Analyse these options, particularly to expose advantages and disadvantages for the environment. Use quantitative methods when these are appropriate. Qualitative evaluation will also be needed. Step 5: select the preferred option. Select the BPEO from the feasible options. The choice will depend on the weight given to the environmental impacts and associated risks, and to the costs involved. Decision-makers should be able to demonstrate that the preferred option does not involve unacceptable consequences for the environment.'

The evaluation of the costs involved, which is required in the implementation of such a policy can be carried out in one of three ways:

1. simple financial analysis, based on conventional costs and revenues in a financial balance sheet;
2. economic analysis, which attempts to incorporate the full social costs and benefits associated with a given option; and
3. comprehensive environmental appraisal, which seeks to encompass monetary valuations derived via economic principles, along with risk-benefit analysis and quantified (but non-monetary) environmental impact analysis.

The reorientation of thinking away from BPM towards BPEO will necessitate turning towards an acceptance, albeit in incremental stages constrained by data and resource cost implications, of the comprehensive environmental appraisal approach to valuation, as explained in the paper by Turner (Chapter 6). Such an approach is also implicit in any move towards integrating environmental costs in the pricing structure of a market economy rather than relying on regulatory limits.

15.2 A SYSTEMS STUDY OF RISKS

To carry out procedures such as these, it will be necessary to generate considerable amounts of data, both on the environmental risks concerned and on the economic costs. What is clearly needed is a 'systems study', in which all steps are evaluated, in quantitative risk terms if possible, and a procedure adopted for assigning weights to the different environmental impacts so that the most favourable can be chosen. The chosen option must be a robust solution, that is, not too dependent on the precise assumptions made, robust against the possibility of accident and maloperation, and robust against changes on a long timescale. Since the different options may well affect different populations, the procedure of choice will have to be seen to be fair. Some of the problems that will emerge are discussed in this paper.

15.3 PCBs

The Royal Commission pose several examples of choices between disposal routes. To show what would be required in a rigorous application of the BPEO concept, let us consider one of them, PCBs. Is it preferable to incinerate or to dispose to secure landfill? It may be quite a topical example.

There is already a considerable literature concerning the hazards concerned with PCBs. They are very stable substances, which persist in the environment; certain isomers are known to be carcinogenic to mice and rats. It is not known for certain whether or not they are carcinogenic to man, but groups of people exposed to high concentrations have developed serious conditions such as necrosis of the liver that have been associated with poisoning due to PCBs or toxic contaminants such as dibenzofurans. PCBs are widespread in the environment at concentrations of 10^{-5} parts per million (p.p.m.), and occur in food in the U.K. at 0.05 p.p.m. (DoE, 1976). Waste contaminated with more than 0.1% of PCBs is regarded as PCB waste. Some relevant factors in a comparative risk assessment would be the following.

1. What is the composition of the waste, 'pure' PCBs or mixed with other material and, if so, what material and in what concentration?
2. Where does the waste arise, how far from a suitable landfill site and from an incinerator? Given the form of the waste, should the risks of transport be considered?
3. Is the landfill site sufficiently characterized to permit acceptance of PCBs? (PCBs at low concentration are accepted in landfill sites. There is

no certain evidence that they are degraded, but they are absorbed on clays and other materials and are not soluble in water.) Is the leachate from the site monitored? What is the long-term fate of the PCB likely to be, if they persist unchanged? What accidents could result in a release at high local concentrations?

4. Are the conditions of incineration suitable? (PCBs should be destroyed at temperatures of $1200 \pm 100°C$, in the presence of excess oxygen.) What risks would arise to the surrounding population due to maloperation of the incinerator? This type of question is a classical 'risk' question. One has to estimate the likelihood of maloperation or malfunctioning occurring and what the products might be. It is known that combustion at too low a temperature can give rise to dioxins and dibenzofurans: how much could be produced? How widely dispersed would these products become, as a function of weather and wind direction? Finally, what would be the pathways to man, and which populations would be at risk? An outline of these calculations is included in the paper by Rae (Chapter 8).

It is very doubtful if the data exists to enable such a quantitative estimate of absolute risks to be made, but a systematic approach would yet be valuable, and might well enable useful semi-quantitative comparisons to be made. At the least, the options could be set out in such a way that informed judgement could be canvassed, using one of the techniques of decision analysis that is now available. And the final judgement would have to take account also of the economic analysis, made by one of the methods listed above.

However, such an exercise is only useful if the decision can be implemented. In the last resort, that depends upon public acceptance that the conclusion is fair, and that any residual risks are tolerably low. We should remember the circumstances in which cargoes containing PCBs were refused entry to the U.K. in the recent past, and sent back to Canada. This was due to the dockers refusal to handle the cargo, perhaps supported by public unease around the incinerators designated to destroy the waste. The critical importance of a sufficient degree of public involvement in decisions on waste disposal is a theme emphasized in later sections of this paper. Public opposition can be so strong as to close out one of the technical options; Backman reports (Chapter 2) that incineration is no longer acceptable in Sweden.

15.4 RADIOACTIVE WASTE DISPOSAL

The disposal of radioactive wastes is a unique case, in that strict quantitative standards of safety are applied. In this case, an absolute calculation of risk is mandatory, in this and other countries, and these calculations have to be carried forward into the far future. It is instructive to ask what differentiates radioactive waste from other categories of toxic wastes, how the calculations of absolute levels of risk are attempted in this case, and what lessons can be learned from the enormous research efforts deployed on radioactive waste disposal that may be applied to other categories.

Forward plans for the disposal of radioactive wastes are described by McInerney (Chapter 12). Several technical reviews are available (Roberts, 1988*b*, 1990) and the regulatory regimes in Europe have been summarized in a recent report from the Select Committee on the European Communities (1988). Wastes are divided into four categories, based on their radioactive content: high-level or heat emitting wastes, the small volume of highly radioactive wastes containing 99% of the fission products and actinides from reprocessing spent nuclear fuel; intermediate-level wastes, which require remote handling and some shielding; and the low-level wastes, which can be handled by standard techniques. There is an even lower category, which can be consigned to ordinary rubbish dumps. The toxicities of the lowest two categories, and of much intermediate-level waste, are comparable with those of other toxic wastes (Flowers, 1985).

Neglecting the very lowest category, there are two distinguishing features which make formal risk assessment a simpler procedure in the case of radioactive than for many toxic wastes. The first is that the volume is much less. Approximate annual amounts of radioactive waste generated in this country are 100 tonnes of high-level, 4000 t of intermediate-level and 25000 t of low-level wastes. This compares with 3.7 million t of hazardous wastes and 1.5 million t of special wastes generated in 1986. The relatively small volumes means that few disposal sites are needed. Two shallow and one deep disposal sites will take all the low- and intermediate-level wastes generated in this country up to 2050, and one deep disposal site for the high-level wastes will eventually be required. The small volumes of intermediate- and high-level wastes also means that expensive packaging is possible, e.g. the vitrification process applied to the high-level wastes. The second significant feature is that the content of toxic material can be closely defined, as it is the radiological hazard that is dominant. The radioactive content of wastes can be recorded and furthermore checked without too much difficulty. Because of the conventions adopted, there is little ambiguity in assigning the wastes to the appropriate category.

The first step in a risk analysis is therefore known, the inventory in a repository. Intermediate- and high-level wastes will be incorporated into durable solid materials (concrete or glass) in corrosion-resistant containers. The only credible path of the radionuclides back to man is through leaching and migration in groundwater. Several barriers to such migration have to be evaluated. The container and the waste matrix constitute the first barrier. The second is the packing material in a repository, and the third is the resistance to flow in the rocks surrounding the repository. Finally, the radiation doses to a population due to any radionuclides reaching the biosphere can be calculated by considering all relevant pathways to man.

Table 15.1 Safety goals in six countries (Parker *et al.*, 1987)

Country	Safety goals (Dose limit, mSv per year)	Time period covered
Federal Republic of Germany	0.3, ALARA	Seeking consensus on 10000 years
France	No decision yet: 1.0 most likely	300 years control of shallow repositories
Sweden	No quantitative limit: ALARA	No time limit
Switzerland	0.1	No time limit
U.K.	0.1 for critical group	No time limit
U.S.A.	0.25, individual	1000 years
	1000 deaths max. over 10000 years	10000 years

Experience so far with these calculations has shown that the most important parameters are the hydrogeological properties of a proposed site, which must be determined by a thorough investigation, including drilling and waterflow measurements.

The criteria of safety that have been imposed on radioactive disposal in different countries are summarized in Table 15.1. They are expressed as a maximum radiation dose to the most exposed population, in this country 0.1 $mSva^{-1}$. It is a very strict criterion and may be compared to the limits from present nuclear installations recommended by the National Radiological Protection Board, of 0.5 $mSva^{-1}$ for members of the public, and 15 $mSva^{-1}$ for workers. A further comparison is with the average background, 2.5 $mSva^{-1}$, and with the variation across the country of 1–100 $mSva^{-1}$, (Hughes *et al.*, 1989).

Several feasibility studies indicate that these very strict safety criteria can actually be met. For example, the EEC sponsored a six-year study of the disposal of vitrified high-level waste in four deep strata, by using named locations where geological data was available (CEC, 1988). The results showed that radiation doses to populations from buried waste will be zero for at least 20000 years and will reach a minor fraction of background levels after some millions of years. A comment on the timescale to which these calculations has been extended seems appropriate. The radioactivity will decrease with time; after about 3.10^4 years, the residual activity is of the same order as that of the original uranium ore (Flowers *et al.*, 1986), and the radiation dose due to buried waste will be less than that from uranium mining wastes (UNSCEAR, 1988). Partly because of these considerations, and since the climatic changes to be expected after 10^4 years will change many of the parameters of risk calculations, a time limit of 10^4 years to these calculations has been set in the U.S.A. and Canada, and probably will be in Germany and France. In other countries, including the U.K., attempts are being made to carry out these formal risk calculations to 10^6 years or even longer, making allowances for the changes in geomorphology and in human societies that may accompany severe climate change such as glaciation (Thompson, 1989). While it is reasonable to assess future situations to be sure that no human disaster is likely to result from waste disposal, it seems unnecessary and perhaps a waste of resources to attempt a formal risk calculation to such low limits in a distant and unknown future (BNES, 1989).

Despite this and other favourable technical risk assessments, progress towards the location of repositories has been slow, because of public unease, the determined opposition of groups opposed to the nuclear power programme and the consequent political difficulties (Roberts, 1988b). From this long history, some lessons can be learned.

1. Despite all the reports of actual radiation levels from the National Radiological Protection Board, the public in the U.K. still put high emphasis on the risks from low doses of radiation, if they arise from industrial activities. This is a special case of the 'low dose' problem, which is referred to below.
2. The concerns of the public do not extend to a million years but to the more immediate future, and include the safety of actual operations, the effect on the economy of the region ('planning blight'), and on the promise of compensation for harm (Kemp and Gerrard, 1988).
3. Continued monitoring and independent reporting is an important feature in winning public confidence.
4. A major programme of public information and public participation is essential in the very early stages of any proposal for a new waste site. Public misgivings are difficult to meet once they are aroused and they cannot be met simply by the application of a strict scientific formalism.

15.5 LAND DISPOSAL OF NON-RADIOACTIVE WASTES

Of the 482 million tonnes a year of the total waste arisings in the U.K., 95% is deposited on land, mainly in landfill sites (RCEP, 1985). Most of this consists of domestic rubbish and organic material, which can generate methane. Properly handled, landfill sites can be a valuable source of fuel; there may be 20–25 large landfill sites in the U.K. that have potential local uses for methane. However, energy recovery does not seem feasible at the majority of sites and, without proper management, the methane can constitute a hazard. HMIP (1989) have estimated that there are 602 active sites with a potential gas problem and 788 closed sites with a potential problem, of which 432 are within 250 metres of housing. The DoE is currently sponsoring a system (HALO) aimed at assisting Waste Disposal Authorities to rank the sites under their control and to instigate remedial or control measures where necessary. Given the variable input and packing of wastes and the influence of temperature, oxygen and water ingress on the generation of methane, a full risk assessment as applied to radioactive wastes can hardly be attempted except on new, highly engineered sites. However, a system of monitoring can probably provide enough information for proper control if the results are analysed in a systematic way.

In 1985, 79% of the 3.5 million tonnes of hazardous wastes were sent to landfill sites. In this regard, methane generation at such sites is only of direct interest in so far as fires could spread toxic contamination. Leachate from the sites that contain toxic material, and the proximity to aquifers, is probably more pertinent. It is also necessary to be aware of possible chemical reactions between the various constituents of wastes, and of the possibility of changes with time.

Two types of landfill site can be distinguished. In one type, water percolation through the site is assumed and the leachate is collected or monitored. In the second type, leaching is inhibited by providing clay caps and impermeable liners. With the emphasis now on containment rather than a 'dilute and disperse' policy, new sites will more probably be of the second rather than the first type. However, if chemically persistent toxic material is added to a landfill site, the potential hazard may not reduce with time and the long-term stability of a site must be considered. For how long can sites be managed and monitored? What is the worst accident that could happen after a site is closed? This type of question has been addressed in the case of the disposal of radioactive waste in near-surface sites and the possibility of inadvertent intrusion into sites containing long-lived activity has also been considered. Past episodes of toxic waste disposal, such as that at Love Canal, U.S.A., show the importance of an assessment of long-term stability and of the possibility of long-term control (Mather, 1989). New regulations in the U.S.A. require disposers to consider landfill disposal of toxic wastes only as a last resort (Fortuna, 1988). That could well be the result of the application of BPEO philosophy in this country.

15.6 SITING OF PROCESS PLANT

In 1985, 1.3% of hazardous wastes were incinerated in the U.K., 3.4% sent for chemical treatment and 3.2% for solidification. With the increasing pressure for more rigorous management of landfill sites, these percentages seem certain to increase, and other papers have illustrated many processes that are being developed. However, processes will only become commercially available if sites can be found for the relevant plant. Since the materials to be handled are, by definition, hazardous, the problems involved in choosing sites are similar to those that have to be faced in the siting of chemical or petrochemical plant which come under the 'Seveso' directive of the EEC, expressed in the CIMAH regulations [1982] in this country.

It would be as part of a choice of process-plus-site as the BPEO for a waste stream that a full environmental impact statement should be provided. This would have to include a statement of the potential hazards to the workforce and the surrounding population that could arise both during normal operations and due to accidents, quantified as far as possible. Full information would have to be provided to the population concerned and emergency plans would have to be prepared.

A typical example, which will show some of the problems likely to arise, is the siting of incinerators. Quite apart from the destruction of toxic material like PCBs, incineration may come to look a more attractive option for the destruction of organic wastes because of the reduction in volume that can be achieved and the consequent lessening of pressure on disposal sites. The U.K. burns only 1.3% of hazardous wastes, and makes less use of incineration of municipal refuse than many other countries (RCEP, 1985). Questions have been raised, however, concerning the safety of the emissions. Dioxins and dibenzofurans can be emitted not only from the incineration of toxic substances such as PCBs but also from the combustion at moderate temperature of plastics such as PVC and lignin, a component of wood.

This type of problem can be tackled by the classical methods of risk analysis, if there is sufficient data available on emissions, since it is emissions that probably pose the greatest hazard. A risk analysis would have to attempt to evaluate all possible pathways to man, and the results might be expressed in terms of possible ingestion, or perhaps as concentrations as a fraction or multiple of background levels.

To go further, one has to assume a relation between risk and dose. Scott (1987) has summarized the risk assessment for the siting of a waste management complex containing an incinerator, together with other treatment and solidification plant and a landfill site. The assessment took into account normal operations, accidents and transport accidents. Exposures were calculated for direct inhalation and from drinking water, crops and livestock. Chronic toxicity was calculated from values for an acceptable daily intake for non carcinogens and a unit cancer risk for carcinogens, using values published by the EPA. The lifetime risk of cancer from exposure at the maximum concentration point was calculated to be 10^{-5} or less, with the highest risk being direct exposure.

The greatest uncertainty in this type of calculation is probably the last step, the estimation of risks arising from low doses. And the greatest difficulty may well be to persuade the surrounding population of the soundness of the estimates. As other papers have demonstrated, public opposition to incinerators may be intense. Both these points need elaboration below.

15.7 RISKS OF LOW DOSES OF TOXIC MATERIALS

The case of dioxins is a good example of a general problem that must be faced in the treatment of many toxic wastes: the difficulty of establishing the risk to humans at low doses. Tschirley (1986) claimed that no ill effects beyond chloracne have been proved in humans, though some dioxins are notably toxic to some animals; the RCEP (1985) accepted that there is no conclusive evidence of lethal harm to humans. Nevertheless, reports of ill health attributed to dioxins formed in incinerators continue to cause much public alarm, such as the well publicised assertion of an excess of congenital eye malformation in children born in the vicinity of the ReChem incinerators in Bonnybridge and Pontypool (Gattrell and Lovett, 1989).

The general method of testing chemicals for carcinogenic or teratogenic properties is to conduct experiments on animals at relatively high doses, and difficulties arise both in the extrapolation to low doses and in the translation of results across species (Batt and Peterson, 1988). Further, the number of new chemicals entering the environment each year is so large that short-term testing is unavoidable. Nor is epidemiological evidence of human harm easy to obtain or to quantify; the best results have been obtained by following up over a long period the medical history of workers exposed to particular substances (Waterhouse, 1988). As one example, the risks from radiation are the most intensively studied of all, and Doll (1990) has summarized the state of current epidemiology applied to the effects of low-level radiation. In spite of all these efforts, the argument about the cause of the small clusters of cases of leukaemia in children found near nuclear installations seems set to continue, though there are reports of similar clusters around non-nuclear sites, and until an explanation is found, the public will remain wary and understandably confused.

The question of what harm may be due to low doses of toxic materials has been termed a 'trans-scientific' question (Weinberg, 1972), in that definitive results are beyond the bounds of classical scientific method, and the answers assumed will depend on a balance of judgement and on the degree of caution deemed to be necessary, which will depend in part on the degree of public awareness and alarm. Regulators cannot honestly aim to define conditions which reduce risks to zero. There is no alternative but to explain a risk oriented approach and to attempt to define tolerable levels of risk, as the Health & Safety Executive are attempting to do for hazardous and nuclear installations (HSE, 1987). What is tolerable will vary from one case to another, dependent on the perceptions of benefits as well as risks, and may require contributions from a wider range of experience than the scientific disciplines. Everest (1990) has discussed how the expert committees which advise government on environmental matters, such as the Advisory Committee on Pesticides, may have to adapt their procedures to meet this type of challenge. Jasanoff (1988) points out that the style of regulatory practice differs from country to country. The risk assessment process used by the EPA in the U.S.A. remains more highly formalized, quantitative and inclined to use conservative assumptions as compared to the more pragmatic arguments from the available evidence that tend to be the practice in Britain or in India. Although Jasanoff considers that U.S. risk assessment policies are moving closer to those of other industrialized countries, these differences of approach will inhibit the emergence of agreed international standards.

Since a guarantee of zero risk can never be given, it is important that any agreement on tolerable levels should be accompanied by an overt mechanism of monitoring suspicious events and of reporting the results. The monitoring would have to be done by an independent agency. The monitoring could extend beyond analytical control of the surroundings of a plant to a collection of health statistics as well, though any such exercise would have to be carried out with great care to avoid confusing effects. The steps that would have to be taken were described by the RCEP (1985), and amount to a considerable effort in any given case.

15.8 THE IMPORTANCE OF PUBLIC PARTICIPATION

The importance of winning public confidence has been mentioned already, but it deserves special emphasis. It is not surprising that waste disposal is an unpopular activity in anyone's backyard. The value of products may be obvious: the value of waste disposal is not. There is an awkward history of mistakes and near-disasters. Hazardous wastes are 'widely perceived as lethal, imposed, beyond individual control, persistent and unfair' (Andrews and Lynn, 1988). In the absence of an agreed basis on which priorities for action can be based, resources tend to be committed to risks which have attracted the most media attention, not to those liable to cause the greatest public harm (Roberts, 1988a). The task of building public confidence in a new and possibly unwelcome operation is bound to take a long time. There is general agreement that full information must be available to the public and that the community must be involved from an early stage. The Royal Commission (1985) go further in recommending that the additional costs involved in monitoring should be borne out of central funds and that local communities affected by waste handling and disposal sites might be offered some discretionary compensation. However, any such scheme will only succeed if the community are first reassured on safety questions, and this requires confidence in the regulatory authorities as well as in the industrial organization concerned.

Fortunately, there are many examples of good relations being built between local communities and industrial organizations, both nuclear and non-nuclear. The recent public consultation exercise carried out by U.K.

Nirex Ltd. has won at least a measure of acceptance in some communities. An elaborate exercise that ended in the successful siting of a waste facility in Alberta, with 79% of the local population being in favour included the following as important points:

1. The perception that the proposal would satisfy a local need;
2. The scale of the exercise in public information, including 59 organizational meetings and 65 community workshops;
3. The key position accorded to the local authorities, who could withdraw if they wished;
4. The care taken to inform the media fully.

In general, the following conditions tend to reduce public distrust: (*a*) the allocation of due weight to social and political factors when developing a 'formula' for assessing risks; (*b*) appropriate involvement of all main affected interests; (*c*) unbiased management of the regulatory process; (*d*) a fair distribution of expertise among affected parties; (*e*) the decisions are not prejudged.

We might note that heavy costs will be involved in such exercises, and also in procedures aimed at determining acceptable levels of risk of potentially toxic materials. Tschirley quotes a source in the Office of Science and Technology, U.S.A., for an estimate of more than a billion dollars spent by the Federal government on research into dioxin toxicity. These costs, where they can be foreseen, should be included in economic estimates if they apply to one option for disposal rather than to another.

Conforming as closely as possible to principles such as those discussed above will place a heavy burden on HMIP and other regulatory authorities. Is the decision-making mechanism equal to the task? It is vital that the public profile of the regulators should be raised and their work seen to be constructive, unbiased and of high standard. The public do not deserve only to be given full information about some accident; they need full information about what is done well, about the philosophy and factual base of regulation and about the procedures that are followed in coming to decisions. These are tasks both for the industries concerned and for the regulatory authorities.

15.9 CONCLUSIONS

From this brief analysis, it can be seen that several subjects have to be addressed before an integrated waste disposal policy for toxic wastes can be in place. The concept of BEPO requires further study in order to establish a basic set of operational principles to guide HMIP. The data requirements, the economic valuation methods and the techniques available for optimization need to be identified. The resource cost implications of an effective integrated inspectorate also require assessment.

The discipline of a systematic risk assessment together with a partial economic analysis can be applied to many situations even before a full theoretical framework is in place. These can be applied to proposals for product substitution, for waste recycling and the segregation of wastes. They can also be applied to the problem of disposal of those waste streams for which no recycling scheme looks economically feasible.

Since wastes exist now and will continue to be generated, it is important that options for disposal are considered objectively against present technology, backed up by sound regulation, and not abandoned because of some history of bad performance. Some specific examples are as follows:

1. The use of landfill should be put on a firmer scientific basis. The monitoring regimes necessary to control gas production and leachate should be determined and implemented.
2. The safety of landfill sites over long time periods requires further study. The use of such sites for the disposal of toxic wastes will depend on estimating both the behaviour of the toxic material with time and the stability of the site in the future.
3. The best means of disposal of persistent toxic materials, particularly metallic compounds and residues should be determined.
4. The use that can legitimately be made of sea disposal should be re-examined. Present policies seem to limit severely any disposal at sea except of inert material. The case for disposal in the deep seas as opposed to disposal in shallow waters should be assessed.
5. Mechanisms of determining acceptable levels of toxic materials in the environment must continue to be developed, and fully explained to the public. Regulation should be supported by monitoring local conditions.
6. Work should continue to determine the safety of incineration, both as a means of destroying stable toxic compounds and as a more general technology of waste treatment.
7. Mechanisms of public participation in deciding on the siting of waste treatment plant should be further explored.

In general, a more positive view of the waste disposal industry, and of the work of the regulatory and advisory bodies, should be inculcated by an enlarged programme of public information and debate.

ACKNOWLEDGEMENTS

The author gratefully acknowledges the value of discussions with Mr R.K. Turner, Mr J.P. Parfitt and Mr S.P. Gerrard of UEA.

REFERENCES

Andrews, R.N.L. & Lynn, F. M. (1988) *Standard handbook of waste treatment and disposal* (ed. H.M. Freeman and W.Y. Lundt), ch. 3.3–3.15. New York: McGraw Hill.

Batt, S.C. and Peterson, P.J. (1988) In *Risk assessment of chemicals in the environment* (ed. M.L. Richardson), pp. 153–174. London: Royal Chemical Society.

BNES (1989) Discussion in *Radioactive waste management* **2**, 129–130. British Nuclear Energy Society, London.

CEC (1988) PAGIS. *EUR 11775 EN*. Brussels: Commission of the European Communities.

CEGB (1988) *BPEO for ash disposed to sea from Blyth and Stella Power Stations. Report CEU/R/88/1*. London: Central Electricity Generating Board.

CIMAH (1984) *The Control of Industrial Major Accident Hazards Regulations.* Statutory Instruments 1984 No. 1902. London: H.M.S.O.

Cook-Mozaraffi, P., Darby, S. and Doll, R. (1989) *Lancet* **ii**, 1145–1147.

Department of the Environment (1976) *Waste management paper* **6**. London: H.M.S.O.

Doll, R. (1990) *Nuclear Energy* **29**, 13–18.

Everest, D.A. (1990) *Science and Public Affairs* **4**, 17–41.

Flowers, R.H. (1985) *Radioactive waste management* British Nuclear Energy Society, London, pp. 109–117.

Flowers, R.H., Roberts, L.E.J. and Tymons, B.J. (1986) *Phil. Trans. R. Soc. Lond.* A **319**, 5–16.

Fortuna, R.C. (1988) *Standard handbook of hazardous waste treatment and disposal*, (ed. H.M. Freeman, N.Y. Lindt), ch. 1.3–1.9. New York: McGraw Hill.

Gattrell, A.C. and Lovett, A.A. (1989) Paper to Institute of British Geographers Annual Conference.

Her Majesty's Inspectorate of Pollution (1989) *First Annual Report* London: H.M.S.O.

Health and Safety Executive (1987) *The tolerability of risk from nuclear power stations.* London: Her Majesty's Stationery Office.

Hughes, J.S., Shaw, K.B. and O'Riordan, M.C. (1989) *Radiation exposure of the U.K. population – 1988 review.* N.R.P.B. – R227. London: H.M.S.O.

Jasanoff, S. (1988) *Risk assessment of chemicals in the environment.* London: Royal Chemical Society. Pp. 92–113.

Kemp, R.V. and Gerrard, S. (1988) *Responses to the way forward.* University of East Anglia: Environmental Risk Assessment Unit.

Mather, J.D. (1989) *J. IWEM* **3**, 31–35.

O'Riordan, R. (1989) *Envir. Conserv.* **16**, 113–122.

Parker, F.L., Kasperson, R.E., Anderson, T.L. and Parker, S.A. (1987) *Technical and socio-political issues in radioactive waste disposal.* Stockholm: The Beijer Institute.

Roberts, L.E.J. (1988*a*) *Risk assessment of chemicals in the environment* London: Royal Chemical Society. Pp. 3–31.

Roberts, L.E.J. (1988*b*) *Proc. R. Inst.* **59**, 259.

Roberts, L.E.J. (1990) *Ann. Rev. nucl. Part. Sci.* **40**, 79–112.

Royal Commission on Environmental Protection (1985) *11th Report*. London: H.M.S.O.

Royal Commission on Environmental Protection (1988) *12th Report*. London: H.M.S.O.

Scott, M.P. (1987) *Waste Management and Research,* **5**, 173–181.

Select Committee on the European Communities (1988) *19th Report, Session 1987–88*. London: H.M.S.O.

Thompson, B.G.J. (1989) *Risk analysis in nuclear waste management* (ed. A. Saltelli *et al.*). London: Her Majesty's Stationery Office.

Tschirley, F.H. (1986) *Scient. Am.* **254**, 21.

United Nations Scientific Committee on Atomic Radiation (UNSCEAR) (1988). New York: United Nations.

Waterhouse, J.A. (1988) In *Risk assessment of chemicals in the environment*, pp. 195–206. London: Royal Chemical Society.

Weinberg, A. (1972) *Minerva* **10**, 209.

16

Towards an industrial ecology

R.A. Frosch and N.E. Gallopoulos

16.1 INTRODUCTION

From earliest times, people have sought knowledge and have created new technologies and industries to improve their life. Because of incomplete understanding of the natural environment, technological innovation and industrialization occasionally have led to unforeseen and sometimes adverse consequences. The ancient Greek myths of Prometheus, who was severely punished for stealing fire from the gods to bring it to humanity; and of Icarus, who plummeted from the sky when the wax with which he glued his wings on melted as he flew too close to the sun, poignantly show the adversities that can result from the quest for knowledge and technology. The feverish industrialization of the 19th and 20th centuries has greatly intensified both the benefits and the ills that are the twin children of technology. The chlorofluorocarbon (CFC) story is a contemporary example of the benefits and risks associated with technological innovation. The CFCs were originally developed to be effective, safe, and non-toxic replacements for the highly toxic and dangerous ammonia and sulphur dioxide refrigerants then in use. At the time of their invention in the late 1920s, CFCs were put through every then conceivable test for toxicity and safety. It is ironic that 50 years later CFCs were found to deplete the stratospheric ozone layer, a phenomenon unknown in the 1920s, which protects humans against harmful ultraviolet radiation.

In our environmentally conscious society, the current vogue is to focus on the ills associated with industrialization and technology: air pollution, water pollution, resource depletion, waste disposal, and urban congestion, top,

The Treatment and Handling of Wastes
Edited by A.D. Bradshaw, Sir Richard Southwood and Sir Frederick Warner
Published in 1992 by Chapman & Hall, London, for The Royal Society
UK ISBN 0 412 39390 5, USA ISBN 0 442 31461 2

but do not exhaust the list. In the rush to find culprits, we seem to forget that industrialization and technology are also responsible for many important improvements such as better nutrition, longer lifespans, more leisure time, and more comfortable habitats. Moreover, technology has been the solution to many environmental and socioeconomic problems caused by industrialization. However, as the world's population multiplies, and as demands for ever-better living standards escalate, the traditional technological approaches to solving the problems of industrialization seem increasingly inadequate. To place the problem in some perspective, consider that by the year 2030, ten billion people are likely to live on the earth and under optimal conditions each human being should enjoy a standard of living equivalent to that of industrial democracies, such as those of Western Europe, the U.S. or Japan. It is doubtful that this goal can be attained unless the problems of natural resource depletion and environmental pollution are solved. Failure to solve them could even plunge humanity into severe decline.

Resource depletion must be analysed in the context of three factors: physical, the total mass of material available; technological, the means available for removing the resource; and economic, the cost for removing the resource. Thus, from a practical standpoint a resource may be considered depleted not because there is so little of it left, but because technology for recovering it economically is not available. Bearing these factors in mind, we can examine the estimated lifetimes of some critical resources such as bauxite (aluminum), copper, cobalt, molybdenum, nickel, the platinum group of metals, petroleum, and coal (resources, reserves, and consumption rates were obtained from: World Reserves, Resources and U.S. Consumption of Minerals: U.S. Department of the Interior Mineral Commodity Summaries 1988; Coal Resources, 1978; 1982 Statistical Yearbook, 1984; Petroleum Reserves and U.S. Consumption: American Petroleum Institute Data Book, 1988; Gallopoulos, 1982). As shown in Table 16.1, at current rates of global consumption, these materials have finite lifetimes of anywhere from decades to hundreds of years. At increased rates of consumption, assuming a world population of ten billion by 2030 and a per head consumption equal to that currently prevalent in the U.S., the lifetimes of some of these materials diminish to less than a decade, unless new resources are discovered, recovery technologies are improved, or substitutes are developed.

To examine fully the dimensions of the environmental pollution problem is beyond the scope of this paper. It can be shown, however, by considering the potential growth in industrial and consumer wastes. In the U.S., the total annual solid waste generated by industry, commerce and consumers is roughly 500 million tons (The garage industry, 1989); this includes 250 million tons of hazardous wastes, which can be either solid or liquid, but excludes agricultural wastes, which are enormous. If in the year 2030 the per head waste generation in the world were equal to the current U.S. per head generation, the world's 10 billion people would generate roughly 20 billion

Table 16.1 Estimated lifetimes of some global resources

	Current consumption rates[a]		Assumed consumption and population for 2030[b]	
	Reserves[c]	Resources[d]	Reserves	Resources
Bauxite	256	805	124	407
Copper	41	277	4	26
Cobalt	109	429	10	40
Molybdenum	67	256	8	33
Nickel	66	163	7	16
Platinum group	225	413	21	39
Coal	206	3226	29	457
Petroleum	35	83	3	7

[a] Years from present, assuming current global consumption rates and present reserves and resources.
[b] Years from 2030, assuming a per head consumption equal to that currently in the U.S., a world population of 10 billion, and present reserves and resources.
[c] Reserves are resources that are known to exist and can be economically recovered with known technology.
[d] Resources include reserves and all other estimated deposits regardless of economics and technology.

tons of waste annually. To get a 'feel' for these numbers, compare them to the current annual world production of either crude oil at 3 billion tons (U.S. Department of Energy, 1988); or of wheat, corn and rice at 1.5 billion tons (The World Almanac and Book of Facts, 1987). It is obvious that we will have to learn how to reach the U.S. standard of living without generating the U.S. equivalent of wastes, which is unusually wasteful as countries go. However, even if waste were generated at a rate typical of a European city such as Hamburg, the worldwide total in 2030 would be close to 10 billion tons annually, which still is an immense amount.

These calculations are not forecasts of the future; they are too simplistic for that, and from an economic viewpoint the availability and lifetime of a resource is a function of its price, and the amount of waste generated is a function of disposal costs – factors that are not included in these calculations. They show, however, the driving forces for recycling, conservation, and the use of alternatives, and they highlight the need for imaginative solutions to these resource and environmental problems. We believe that these solutions can be found within the notion that the industrial system ought to be modified so as to mimic the natural ecosystem in its overall operation. In other words, the industrial system should not be viewed or treated as a collection of individual and virtually independent manufacturing and other operations in which raw materials and energy are transformed into products to be sold and waste to be disposed of separately, but rather as an interacting web of inputs, processes, and wastes, all to be thought of together, as discussed by Bradshaw (Chapter 1).

16.2 THE INDUSTRIAL ECOSYSTEM CONCEPT

Industrial processes consist of a set of technologies by which materials and energy are transformed into products that people want and use. By technologies we do not mean devices or gadgets, which are end products, but rather the understanding required to make products, based either on an understanding of nature and natural law or upon phenomenologically understood crafts. At the start of the process, raw materials are introduced, which may have been extracted from the earth, grown or harvested from living things, or may themselves have been manufactured products. At the end, each product that has gone to a consumer may eventually be discarded by the user, at which point it may become a disposal problem or, as scrap, an input to some industrial process. This generalized scheme is shown in Fig. 16.1. Conventionally, the industrial system is treated as an assemblage of many such individual processes that frequently operate as if they were independent of each other.

Ayres (1989) likened the industrial transformation of materials to that of the biosphere, and coined the phrase 'industrial metabolism' to emphasize the importance of biological-type behaviour in industrial processes. Ayres' concept is an important contribution toward improving the industrial system. However, we believe that metabolism, the sum of the physical and chemical processes in an organism by which its material substance is produced, maintained and destroyed, and by which energy is made available, does not fully describe the analogue required for the ultimate improvement of the industrial system. Our preferred approach is to view the industrial network as an 'industrial ecosystem', analogous in its functioning to a community of biological organisms and their environment. Some of the organisms use sunlight, water, and minerals to grow, while others consume the first, alive or dead, along with minerals and gases, and produce wastes of their own. These wastes are in turn food for other organisms, some of which may convert the wastes into the minerals used by the primary producers, and some of which consume each other in a complex network of processes in which everything produced is used by some organism for its own metabolism. Similarly, in the industrial ecosystem, each process and network of processes must be viewed as a dependent and interrelated part of a larger whole. The analogy between the industrial ecosystem concept and the biological ecosystem is not perfect, but much could be gained if the industrial system were to mimic the best features of the biological analogue. An ideal industrial ecosystem may never be attained, but even to start approaching the ideal both manufacturers and consumers will need to change their habits.

If the manufacturing web is viewed as an industrial ecosystem, then it becomes important to choose industrial processes that, like their counterparts in the biological world, operate at high efficiency diversity and resiliency.

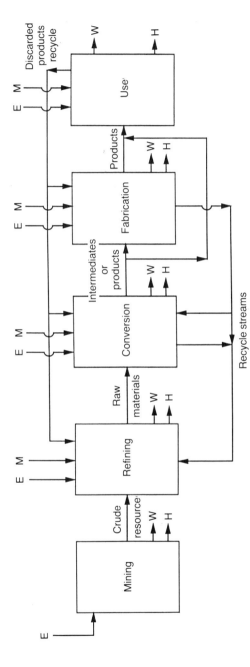

Figure 16.1 Generalized manufacturing and product use subsystem. (E, energy input (direct or indirect such as fuel for transportation); M, additional materials (raw, intermediate or finished) required to process or use the primary material; W, waste material; H, waste energy.)

Thus, energy requirements should be minimized, as should waste generation and consumption of increasingly scarce materials. Moreover, industrial wastes, discarded products, and trash and garbage should be utilized as inputs to various industrial processes in different industries, in a way analogous to the cycling of nutrients by various organisms in an ecological food web. Finally, the system should be diverse and resilient enough to absorb and recover from unexpected and severe shocks or surprises.

The industrial ecosystems approach to manufacturing begins with individual process optimization, a technique of engineering analysis that balances the design of a manufacturing process against the needs of the customer to obtain the best possible set of operating characteristics for a given job. It is a way of balancing conflicting requirements for energy and materials in the specific internal steps of each process. As is shown later, in most industrial process optimization, environmental impacts are considered final external effects at the output end, to be mitigated by treatment or disposal, rather than intrinsic internal parts of the total system. In addition, optimization of a subsystem or a process component does not necessarily yield a system that is optimized overall. This raises the very important and difficult-to-answer question of how large the optimization domain should be. An added complication is presented by the potential conflict between optimization from a mass or thermodynamic sense, and optimization from an economic or environmental or societal sense. Although we offer no specific solutions to these problems, our discussion of the industrial ecosystem provides some of the framework within which these problems should be handled.

An ideal industrial ecosystem does not exist today, and even individual manufacturing processes are less than perfect. Furthermore, little consideration is given in the design of industrial products and processes to whether variations in the design of the products and manufacturing processes might mean that waste materials could be minimized or used elsewhere in the industrial system. Such consideration is sometimes given to the use of 'designed offal' ('engineered scrap') of metals and some plastics and in the fabrication of new products from materials previously considered as wastes. Also, the use of the waste heat from electric energy generation for process heat (cogeneration) is increasingly practised. However, such considerations are not general except where there are obvious immediate economic advantages within the plant, or within the company itself, and effects on diversity and resiliency are essentially ignored. It is complicated and difficult to go beyond the current boundaries of designing industrial processes and products to minimize wastes and pollution, and the economic incentives to do so are rarely there.

These difficulties, along with the bright spots, will be explored by examining three subsystems of the existing vast manufacturing network. These three subsystems (iron, plastics and platinum-group metals) have evolved to differing degrees of material and energy closure and waste control. They are not good microcosmic models of the ideal industrial ecosystem, but examining them should provide insight into what is lacking and how the ecosystem approach could help. The conversion of iron ore to steel and other products represents a very mature process with a several thousand-year history, although extensive steel production did not start until the 19th century. The plastics-from-petroleum cycle is less mature than the iron cycle, being less than a hundred years old, since the first all-synthetic plastic (Bakelite) was introduced in the first decade of the 20th century. Two aspects of the even younger platinum group-metal-to-catalyst cycle provide a third example; platinum-group metal catalysts were used in industry as early as 1940, but their intense use dates only to the early 1950s.

16.3 THE IRON SUBSYSTEM

All economies use large amounts of elemental iron, the predominant component of steel and cast iron; the principal compositional difference between them is that steel has a much lower carbon concentration. In the U.S., iron ore is usually mined in huge open pit mines with depths of up to several hundred feet. The iron ore is concentrated and pelletized on-site at the mine. The concentrated ore is heated in a blast furnace together with coke, limestone, and air (oxygen) to produce pig iron, which is the feedstock for cast iron and for steel.

Along each step of the steelmaking process (Fig. 16.2), by-products,

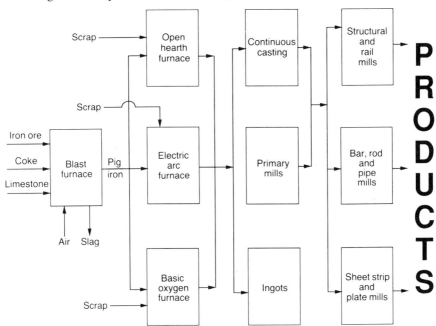

Figure 16.2 Steelmaking.

wastes, and pollutants are generated (Lankford *et al.*, 1985). For the mature iron cycle, these are generally minimal, due to extensive recycling or alternative use of the by-products. The blast furnace is an excellent example of high energy- and material-use efficiency. The slag produced in the blast furnace, as well as in other parts of steelmaking is used to manufacture aggregates for road paving, cement, concrete products, filter media, and many other products. Some steelmaking slags are also recycled to the blast furnace because of their high iron content. Even those by-products or wastes and pollutants that cannot be economically recovered or reused are tightly controlled through either voluntary efforts by the steel industry and/or governmental regulations.

Because iron is widely used and its ferromagnetism facilitates identification and separation, its recycling is extensively practiced (Fig. 16.3). Enormous amounts of scrap, generated within the steel and iron mills and by consumers, join iron ore to produce iron and steel products (Anon. 1985). The offal generated by stamping steel parts for automobiles (roofs, doors, fenders, etc.) is used as material for recycling into engine blocks and other castings. For example, all four General Motors foundries use exclusively their own foundry scrap and steel scrap from other General Motors operations to make all their products. However, the overall iron cycle is not completely closed, mainly because of incomplete recovery of scrap iron from discarded consumer products. Much of this scrap is scattered around the countryside, it is often considered a blight rather than an asset, and corrodes away a little every year. The iron in some discarded consumer products is lost forever because it is dispersed in concentrations far lower than those found in ore, e.g. steel cans scattered among unprocessed municipal wastes.

Production and consumption of iron ore, steel, and iron and steel castings in the U.S. during 1982–1987 was the lowest it had been for any six-year period since the end of World War II (American Iron and Steel Institution (1980–87) and Phoenix Quarterly, 1984). However, the amount of scrap has continued to increase, reaching about 800-million tons in 1987 (American Iron and Steel Institute). Much of this increase has accumulated in these past six years of low iron and steel consumption. A number of factors have contributed to this unwelcome development.

As a result of efforts to reduce air pollution and increase efficiency, the basic oxygen furnace (BOF) started replacing the open-hearth furnace as the primary steel-industry furnace in 1958, and the latter now accounts for less than 3% of production. This decreased the demand for scrap because the BOF uses less scrap than the open hearth furnace. At the same time, the electric furnace, which uses virtually 100% steel scrap, increased its share of steelmaking, but the increase in scrap consumption was far from enough to offset the decrease caused by the switch to the BOF. Together, the decrease in demand for iron and steel and the switch to the BOF significantly decreased

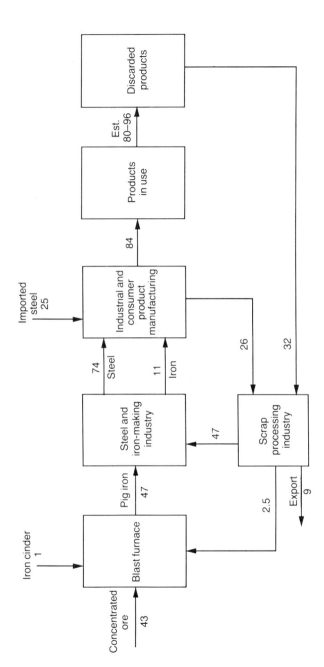

Figure 16.3 The iron cycle in the U.S. – 1984 data. (Note: the numbers are annual flows of material in millions of tonnes. Inputs and outputs do not always balance as fuel and slag are not accounted for.)

the amount of iron scrap reprocessed. Meanwhile, an economic mechanism was lacking to account for the adverse environmental impacts of the accumulation of scrap material, or the possible very long-term effects of resource depletion.

16.4 THE PETROLEUM–PLASTICS SUBSYSTEM

For separation and identification purposes, the chemical complexity of iron products is mitigated by the fact that iron is magnetic and it is the predominant elemental component of these products. Plastics, on the other hand, are a variety of complex organic compounds, they lack properties conducive to simple identification, and several of them can be found in a single consumer product. Consequently, recycling of plastics is difficult because of collection and separation problems. Also, returning plastics to their original organic and inorganic constituents is often technologically infeasible, or economically unattractive.

Because of their durability and visibility in wastes, plastics are considered a serious environmental problem, a reputation that may not be fully justified. In fact, plastics replaced other materials because of advantages related to energy and material conservation, and because of desirable properties for their intended use. Using a 'total-systems approach', the Midwest Research Institute (MRI) (Hunt and Welch, 1974) compared the energy, material, and environmental characteristics of several plastics with those of the materials they displaced in various food containers. Results for one of the plastics, polyvinyl chloride (PVC), are shown in Table 16.2. For the same number of bottles produced, PVC consumes much less total material and energy than glass; also, PVC bottles are responsible for less solid waste and less atmospheric and waterborne emissions than the glass bottles they replaced. Results for other plastics replacing paper, aluminum, or steel were similar to those for PVC. However, accounting for total energy and mass of materials and

Table 16.2 Comparison of PVC and glass in the manufacture of half-gallon bottles (basis of calculation: 1 million bottles)

	PVC	*Glass*
Raw materials[a] (lb)	200426	3919809
Energy (million Btu)	12177	25739
Water (thousand gal)	2007	6981
Solid wastes (cu ft)	965	17279
Atmospheric emissions (lb)	57363	126755
Waterborne wastes (lb)	8914	14337
Post-consumer solid wastes (cu ft)	5317	15452

[a] Crude oil and natural gas raw materials are included in the energy category.

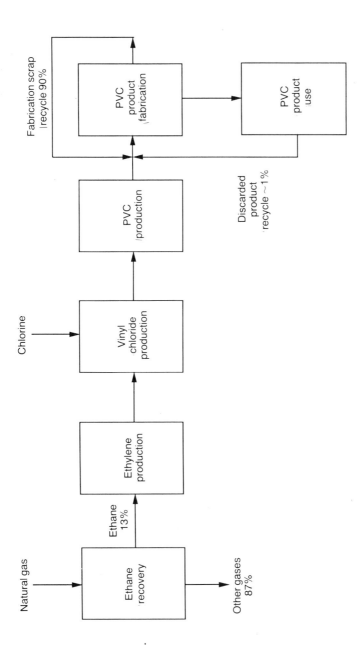

Figure 16.4 The polyvinyl chloride (PVC) cycle. (Note: for simplicity, recycling steps in the production processes have been omitted.)

wastes does not necessarily assess the total environmental impact of a process. Environmental impacts must be accounted for explicitly because they may not be directly proportional to the energy or mass fluxes. The MRI study weighted the environmental impacts and concluded that plastics were still superior to the replaced materials.

Plastics also save transportation fuel by decreasing the weight not only of the products that use them (for example, plastic instead of glass milk bottles), but also of the cars, trucks, aircraft, and other transportation vehicles that are used for the conveyance of goods and people. In addition, products made of plastics are frequently safer to use. Just consider the dramatic drop in household injuries, whether minor cuts or major lacerations, that resulted from the replacement of glass by plastic bottles in kitchens and bathrooms for such products as milk, soaps, and cleaners.

To show both problems and progress with plastics in terms of the manufacturing cycle involved, we selected PVC, which is probably the most environmentally challenging of the plastics both in manufacturing and recycling. Using the most difficult of the plastics as an example, it can be concluded to show that plastics do not fully deserve their reputation as environmental villains, and we show the use of the ecological concept in examining the environmental impacts of plastics.

Focus is on the major processes and reuse factors in manufacturing and products, rather than on the problems of plastics as distributed trash on land and at sea, because these latter problems would be greatly decreased or eliminated if an economical industrial ecosytem comes increasingly into practice.

A hydrocarbon and inorganic chlorine are reacted to produce PVC. The presence of chlorine makes the environmental impact of this material potentially greater than that associated with a plastic such as polyethylene, which is composed entirely of hydrocarbon units. The PVC subsystem is shown in Fig. 16.4 (Nass and Heiberger, 1986). Typically in the U.S., the feedstock for PVC manufacture is natural gas. Elsewhere, the feedback is naphtha, a petroleum fraction. Ethylene is produced from either hydrocarbon feedstock, and it is chlorinated to the monomer vinyl chloride, which subsequently is polymerized to produce various PVC plastics. The energy and material consumption of the overall process has been minimized through many improvements. For example, the U.S. has been changing over to naphtha as the feedstock to conserve energy. Energy conservation is also the reason for the use of membrane cells in the electrolytic process in which chlorine is made from sodium chloride. An added advantage is that the membrane cells do not use the hazardous materials (asbestos and mercury) of previous processes.

As soon as the toxicity and carcinogenicity of vinyl chloride monomer were recognized, its emissions during manufacturing were stringently controlled, but in the classic end-of-stream way. As shown in Fig. 16.5, the

Table 16.3 Major uses of polyvinyl chloride

Use	Percentage
Pipe and fittings	35
Films and sheet	15
Flooring materials	10
Wire and cable insulation	5
Automotive parts	5
Adhesives and coatings	5
Other	25

unreacted vinyl chloride is removed from the finished PVC and from waste waters by stripping, usually with steam. The vinyl chloride from the strippers and the monomer recovery unit is destroyed in an incinerator equipped with scrubbers to remove hydrochloric acid from the exhaust (Nass and Heiberger, 1986).

As shown in Table 16.3, of several uses of PVC, pipe and fittings are the largest ones, but no single use exceeds 50% of production (Rider, 1981). From a recycling viewpoint, having such a distribution is both a disadvantage and an advantage: since the material is dispersed widely through the economy, collection and recovery are hampered; on the other hand, the material may be recycled into uses that may not have as strict requirements for material properties or absence of contaminants as does the use from which the material orginated. The question of contaminants is particularly important in the reprocessing of plastics. The temperatures at which many plastics melt are too low to ensure sterilization; therefore, some recycled plastic is unacceptable for food containers where the plastic directly contacts the food. This is very different from steel and aluminum, which melt at sufficiently high temperatures to ensure that food containers manufactured from the reprocessed materials are sterilized in the manufacturing process.

For many discarded plastics, three recycling alternatives exist: they can be cleaned and ground for reuse in a moulding or other process; they can be chemically converted back into the original starting materials and reused as feedstock; they can be an energy source if burned in an incinerator or other combustion device equipped to recover heat. For PVC, chemical recycling is not a viable option, but reuse and incineration are practised.

It is believed that 90% or more of the uncontaminated PVC scrap generated in manufacturing processes is reused (Chemical and Engineering News, 1989; Anon., 1989*b*). Scrap generated while trimming extruded and molded PVC parts in General Motors plants is segregated by color, reground, melted and used along with virgin PVC. The reuse of contaminated PVC, PVC used and discarded by consumers or contaminated as a result of

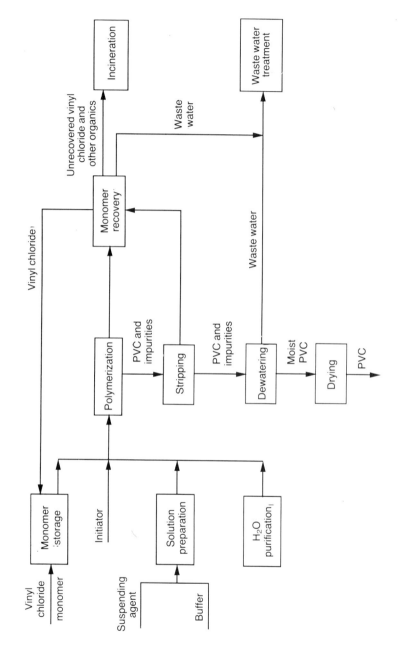

Figure 16.5 Polyvinyl chloride (PVC) manufacture by the suspension polymerization process.

processing, is much smaller, about 1% (Anon., 1989*a*). The contaminated plastic can be reused in products that have lower purity requirements than those of the original use; for example, PVC bottles, that for various reasons cannot be reprocessed into new bottles, that used to make PVC drainage pipe and molded packaging materials. Considerable progress is being made in recycling specific scrapped PVC products, such as water bottles.

Because of its energy content (equivalent to that of wood or paper), scrapped PVC can be incinerated, but the presence of chlorine poses problems. In incinerators, the chlorine in the PVC is converted to hydrochloric acid, which if not removed from stack gases can contribute to acid rain. Also the chlorine can produce small amounts of the carcinogenic dioxins. Consequently, the incineration of PVC has been discouraged. Recent tests show that neither hydrogen chloride nor dioxins are a problem when incinerators are designed and operated properly (Anon., 1988) but environmentalists and regulators are not convinced. On the other hand, PVC producers are confident that PVC can be incinerated safely. As a result of these pressures, PVC use in a variety of products is threatened by other 'environmentally clean' plastics that have physical properties similar to it, but contain no chlorine.

In contrast to PVC, such plastics as polyethylene (PE), polypropylene (PP), and polyethylene terephthalate (PET) face fewer obstacles for incineration or recycling. They are preferred as incinerator feed because they do not contain chlorine or other environmentally harmful elements, and PE and PP have about 50% more energy content per unit mass than PVC. The recycling of PET, which is extensively used in carbonated beverage bottles, seems to be a success story in the U.S., largely helped by the deposit laws in nine states (California, Connecticut, Delaware, Maine, Massachusetts, Michigan, New York, Oregon, and Vermont). In those states, about 80 to 95% of the bottles sold are collected and recycled; nationwide, the recycling rate for beverage bottle PET is about 20%. Recycling of PET is interesting in that it exemplifies two approaches to the problem. Reclaimed PET can be ground and then mechanically reprocessed (molding, extrusion) to new products, such as fibres for fibrefill and carpets. Wellman, Inc. of Clark, New Jersey, is the largest recycler of PET for these purposes. The PET can also be reclaimed through chemical means, in which case the PET can be converted either to polyols used in the making of polyurethane foam insulation, or to unsaturated polyester resins used in bathtubs and the like, or to the original starting materials for repolymerization to PET.

In the future, the recycling of mixed plastics may be possible. At least one company, DuPont, is developing new types of mixed-plastics products based on the use of proprietary additives to improve the compatibility of diverse post-consumer plastics for recycling. Such a development would greatly enhance the potential for the recycling of plastics and should alleviate problems of plastics disposal. However, plastics recycling also has its critics, who accurately observe that recycling only delays the conversion of plastics into waste. Consequently, the ultimate solution to the plastics waste problem has to be incineration for energy recovery, or the decomposition of the plastics to their starting materials whenever feasible.

16.5 THE PLATINUM GROUP METALS SUBSYSTEM

The last example involves the platinum group metals (PGM), platinum, palladium, rhodium, ruthenium, iridium, and osmium. The largest deposits of PGM are found in the Republic of South Africa; smaller deposits are in the Soviet Union. Huge amounts of ore have to be removed from shaft mines as deep as 3000 feet underground because the concentration of PGM in the ore is very low, on the order of 7 p.p.m. (Loebenstein, 1985). In South Africa, approximately 20-million metric tonnes per year of ore must be mined to produced a mere 143 metric tonnes of purified PGM. As shown in Fig. 16.6, about 60% of this PGM is fabricated into various metal products, including jewelry (Robson and Smith, 1988). Ultimate recycling of these products is almost 100%. Another 40% of the PGM extracted from the ore is converted into chemicals and catalysts for use in the chemical, petroleum, and automotive industries. In the petroleum and chemical industries, the PGM are used mainly as process catalysts, whereas in the automotive industry they are used to control the emission of pollutants from individual automobiles. In composition, automotive and petrolem refinining catalysts

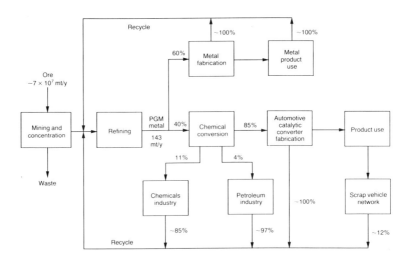

Figure 16.6 The platinum group metals (PGM) cycle. (Note: ore and PGM numbers are for South African production.)

are very similar: they consist of an alumina-based carrier over which very small amounts of one or more of the PGM are dispersed. In recycling, however, the two cases are vastly different. In the chemical and petroleum industries, where the use of the PGM is concentrated to a few manufacturers in a few locations, 85% and 97%, respectively, of the PGM is recovered and recycled (Loebenstein, 1985).

In contrast, only about 12% of the PGM is recovered from automotive catalysts (Robson and Smith, 1988) which until very recently were used only in the U.S. and Japan. Automobiles are consumer products with average lifetimes of a decade or so, thus keeping the catalyst out of the recycling market for a long time; in the petroleum and chemical industries, the catalyst regeneration and recovery cycles can be as short as a few months. Furthermore, the use of automotive catalysts is only fifteen years old, and the amount of PGM in each automobile catalytic converter is small (about 0.05 troy ounces for platinum); not until recently did the recycling industry have a reasonably high and constant feed volume of PGM. Perhaps the most important constraint in recycling automotive catalysts is that literally millions of individual catalytic converters in the U.S. are dispersed among many thousands of automotive scrap yards and over 1800 automotive scrap recyclers; this has unattractive economics and logistics implications for locating, collecting, and emptying the catalytic converters, and for transporting the catalyst material to a reprocessing plant. Estimates show that unless the price of platinum exceeds $500 per troy ounce, recycling of automotive catalysts is uneconomic. This is primarily due to collection difficulties, as several good processes exist for recovering the PGM from catalytic converters. For example, Texas Gulf Minerals and Metals uses a pyrometallurgical process designed specifically to recover platinum, palladium, and rhodium from automotive catalysts. Their plant started operations in September, 1984, and the refined platinum, palladium, and rhodium yields are approximately 90%, 90%, and 80%, respectively.

Since 1987, some Japanese companies (Nippon Engelhard, for example) have set up collecting organizations in the U.S. to acquire automotive catalysts for reprocessing in Japan. This development is likely to increase prices for used automotive catalysts, making their recycling more attractive. The forthcoming use of catalysts for emission control on European vehicles will increase demand for PGM, and should increase the PGM price, thus improving recycling prospects. In the future, a considerable increase in automotive catalyst recycling from the current levels is likely. However, because of the great dispersion of the catalytic converters and associated logistic and economic factors, it is unlikely that recycling of automotive catalysts will ever approach the 85%–97% rate of chemical and petroleum refining catalysts.

16.6 POLLUTION AND WASTES CONSIDERATIONS

The iron, PVC, and PGM subsystems exemplify the differing extents to which individual industrial subsystems approximate the ideal closed system or mimic individual biological organisms or specific biological subsystems. This exposition, however, has not sufficiently considered two important aspects of the ecosystems approach: the role of by-products and wastes found in voluminous streams of process effluents from these subsystems and the consumer side of the problem. A discussion of these topics is needed before discussing how to approach the ideal industrial ecosystem.

An example of pollutants emitted as miniscule components of effluent streams is the hydrocarbon (HC) emissions from the various steelmaking operations such as the blast and steelmaking furnaces, the semifinishing mills, the finishing mills, and the product manufacturing plants. Hydrocarbons are also emitted in the manufacturing of plastics, and many other products, as well as from both stationary and mobile combustion plants. The HC concentrations in the effluents are very small, but small concentrations of HCs in the atmosphere are undesirable because they lead to the creation of ozone in the lower troposphere. Likewise, chlorinated materials are emitted in making steel or plastics and in refining the PGM because hydrochloric acid is used in some of the manufacturing steps. As mentioned earlier, chlorine compounds can contribute to acid rain and other environmental problems.

From an environmental impact aspect, the emission of very small amounts of materials as wastes or by-products from industrial processes is a most difficult problem to solve: internal process economic incentives usually do not exist for the manufacturer to capture and treat these materials to make them environmentally neutral, or to shift to processes with more benign, or more easily handled effluents. The concern generally has been translated into governmental regulations requiring the manufacturer to treat the effluent to remove the hazardous components. Most frequently, the governmental rules and regulations have been expressed as limits on specific effluent rates or concentrations that the manufacturer is allowed. Companies must meet regulatory requirements, but there are no direct economic advantages for manufacturers who capture and treat low-level effluents or who shift to production processes with more benign by-products.

By-products and effluents created during manufacturing represent only the supply side of the industrial ecosystem. The demand side is the consumer, who takes in manufactured goods and produces scrap that could be the raw materials for the next cycle of production. For the industrial–ecosystem approach to become widespread, changes in manufacturing must be matched by changes in consumers' demand patterns and in the treatment of materials once they have been purchased and used.

The behaviour of consumers in the U.S. today constitutes an aberration in both time and space. Whereas a typical New Yorker, for example, discards nearly two kilograms of solid waste every day, a resident of Hamburg or Rome throws out only about half that (Worldwatch Institute, 1987), as New Yorkers did at the turn of the century. Moreover, U.S. consumer habits and waste-management practices form a complex pattern that hinders efforts to reduce waste generation and the growing pressure on municipal landfills. The bulk of consumer wastes consists of organic materials (Chemical and Engineering News, 1989) that could relatively easily be composted, recycled or burned to produce energy, but instead are stored in landfills, for which land was readily available in the past and where costs were low.

Today, as landfills across the U.S. near capacity, many communities have initiated garbage-sorting programmes to reduce the amount of unrecycled waste; more initiatives are likely to follow. Some other countries have already instituted fairly sophisticated collection and treatment practices that go well beyond standard sorting and recycling. Japan, Sweden and Switzerland, for example, have set up collection centres for batteries from portable radios and other consumer products. The batteries contain heavy metals that render composted wastes unsuitable for fertilizing crops; the metals also contaminate fly and bottom ash from incinerators, so that the ash must be disposed of as hazardous waste.

An effective infrastructure for collecting and segregating various consumer wastes can dramatically improve the efficiency of the industrial system. The American and other consumers may have to stop generating huge volumes of unsorted wastes, an action that may actually enhance rather than diminish their living standards as a whole. However, most attempts to introduce the sorting of wastes, or even recycling in general, have tended to rely on exhortation rather than inducement, and there are apparently few cases in which an infrastructure to make waste sorting and recycling simple and straightforward has been provided before regulation. Moreover, landfills for municipal wastes are running out of space as rapidly as are those for industrial waste; consumers will soon find themselves facing the same economic incentives for waste reduction that producers face today for hazardous wastes. The present system of paying for waste management through undifferentiated taxes hides the true costs of disposal from the consumer and provides no incentives for better waste management on the consumer's end. A system that exposes consumers directly to their disposal and other environmental costs, forces them to 'pay-as-you-go', and provides a reasonable infrastructure to make recycling convenient, would facilitate progress to a more rational environmental protection system by providing some direct economic feedback of the consequences of waste.

16.7 APPROACHING AN IDEAL INDUSTRIAL ECOSYSTEM

Process and subsystem optimization and changes in consumer behavior to minimize pollution and wastes would go a long way toward improving the present system, but these measures alone would still leave us short of the ideal industrial ecosystem, which requires the integration and coordination of industrial processes and subsystems to closely resemble the functioning of biological ecosystems. To achieve this integration, improved economic, educational, and societal structures will be required.

Conventional economic methods take into account only the immediate effects of production decisions. If a manufacturer produces non-recyclable containers, for example, taxpayers at large bear the increased costs for collecting and dispersing them; if a powerplant reduces emissions that cause acid rain, communities elsewhere are likely to reap the benefits. Financial returns to the manufacturer or utility are generally indirect. In most cases, and as already discussed in connection with consumer wastes, the costs are passed on to taxpayers in undifferentiated taxes so that the consumer is also unaware of the true cost of waste disposal, and consequently also lacks the economic driving force to alter product use and disposal patterns. Traditional manufacturing processes are designed to maximize the immediate benefits to the manufacturer and the consumer of individual products in the economy rather than to the economy as a whole. A holistic approach will be required if the proper balance between narrowly defined economic benefits and environmental needs is to be achieved. Odum (1989) has also discussed the holistic approach to solving environmental problems, and the deficiencies of the 'piece-meal' measures to solve complex problems.

Instead of absolute or command-and-control rules, some economists have long advocated financial incentives to reduce pollution. These include investment or research credits, tax relief, or fees or taxes imposed on manufacturers according to the amount and nature of the hazardous materials they produce. Such measures can help pay for treatment or disposal; more important, they give companies an incentive to change their manufacturing processes so as to reduce hazardous-waste production. Fees and taxes for pollution make environmental costs internal, so that they can be taken into account when making production decisions. Recent, extensive discussions of this topic have been published by Ruckelshaus (1989) and Pearce *et al.* (1989).

Pollution fees have come under fire from environmentalists and industrialists as 'licenses to pollute' and as 'distortions of the market', respectively. Both criticisms are potentially valid. Companies can treat fees that are too low as a cost of doing business and pass them on to customers rather than changing their operations; fees that are too high may force companies to reduce emissions of specific pollutants without regard to other environmental effects or to financial burdens.

Suitably set charges or incentives, however, can be an effective means for manufacturers to incorporate societal costs of pollution and waste into their cost-accounting systems. In the U.S., rising landfill fees are already causing manufacturers to curb hazardous waste generation. Such a cost feedback for other pollutants could make it more attractive to solve problems at the source rather than to destroy or dispose of effluents once they have been created. Such fees enable manufacturers to share in the overall economic saving accruing from reduced levels of hazardous materials. Providing economic incentives would harness manufacturers' strong competitive drive to reduce costs; indeed, manufacturers who ignore this imperative perish from the marketplace. Such an allocation would also help consumers change their product use and disposal patterns, since pollution control costs would necessarily be reflected in product prices.

In advocating the use of economic incentives, we do not deny the importance of 'free-market' forces. On the contrary, the economic incentives are to supplement the free-market forces, which in many cases are already working well. We would like to find more ways to internalize societal (environmental) costs so that economic market forces will lead to natural solutions to waste and materials problems. We also would like to avoid regulatory 'solutions' which prevent 'ecological' solutions. Such regulatory problems can include bureaucratic certifications for the transportation of materials that are sufficiently onerous and expensive to effectively prevent using wastes from a process in one location as production inputs to a process in a different location.

As an example of the free-market approach, consider Meridian National in Ohio, a midwestern steel-processing company; it reprocesses the sulphuric acid with which it removes scale from steel sheets and slabs, reuses the acid and sells ferrous sulphate compounds to magnetic-tape manufacturers. Also at ARCO's Los Angeles refinery complex, a series of relatively low-cost changes have reduced waste volumes from about 12 000 a year during the early 1980s to about 3400 today, generating revenue and saving roughly $2 million a year in disposal costs. The company sells its spent alumina catalysts to Allied Chemical and its spent silica catalysts to cement makers. Previously these materials were classified as hazardous wastes and had to be disposed of in landfills at a cost of perhaps $300 a ton.

Alkaline carbonate sludge from a water-softening operation at the ARCO refinery goes to a sulphuric acid manufacturer a few miles away, where it neutralizes acidic wastewater. (The acid manufacturer previously purchased pure sodium hydroxide for the same purpose). A few outflow pipes have been rerouted to improve access for loading, and plant personnel must track the pH of their sludge, but the total investment has been minimal.

ARCO's situation is not unique; other major refiners and chemical manufacturers are engaged in similar efforts. For example, investments of $300 000 in process changes and recovery equipment at Ciba-Geigy's Toms River plant in New Jersey reduced disposal costs by more than $1.8 million between 1985 and 1988. Dow chemical established a separate unit to recover excess hydrochloric acid, which it then either recycles to acid-using processes or sells on the open market. The operation recovers a million tonnes of acid a year at a profit of £20 million.

To usher in the ecosystems industrial phase, changes in the economic structure will have to be accompanied by the strong presence of the concepts of a holistic or ecological approach and of system optimization in engineering and technological education. These concepts are either not taught or are taught in such a limited way as to have a very small impact on the problems involved with the effects of manufacturing processes on the environment (Friedlander, 1989). However, changing the content of technological education alone will not be enough. These concepts must be instilled into the practices of government and industry, into our social ethos, and they must be recognized and valued by the communications media. Industry could take the lead to help usher in the industrial ecosystem approach by establishing inter-industry organizations to facilitate it. Such an organization has already been proposed by Allenby (1989) to cope with the challenges presented by the potential for global warming due to the release of carbon dioxide and other trace gases into the atmosphere. He suggested the formation, with the support and encouragement of government and international organizations, of a World Industry Sustainable Technology Institute as 'the vehicle for contributing the considerable resources and expertise of industry to resolve the problems of global climate change without the shortcomings that existing industrial organizations might bring to this function'.

While the industrial ecosystem approach is being shaped into a practical tool, governmental regulation at the local, national and international level will be needed to accomplish some of our environmental and societal goals. However, such regulation can play a useful role only when it is based on sound technology, is altered as technology changes, and is cast in ways that encourage, or at least do not discourage, the development of alternative industrial processes. However, such a wise regulatory framework will not be achieved unless the present adversarial relations between government, industry, and environmental groups are replaced with cooperative efforts. In particular, the technological know-how of industry must be fully integrated into control efforts to avoid counterproductive control measures and costs that are not in line with the benefits derived from the control measures. Furthermore, adoption of economic incentives as part of the regulatory framework would accelerate the transition to the ideal industrial ecosystem.

Human beings, individually or collectively, do not have the knowledge and reasoning capacity required to completely and satisfactorily solve these very complex problems. But we are tenacious enough to keep on trying, and smart enough to be ready for surprises, both good and bad, as we face the future. What we are seeking is a better system for the coordination of technology, industrial processes, and consumer behaviour that has the following characteristics: energy and material use is optimized; wastes and pollution are minimized, and there is either an economically viable use or an environmentally sound disposal method for every end-product, and for every waste that cannot be economically eliminated by a change in manufacturing processes. Furthermore, the ecosystems approach would safeguard the diversity and improve the resiliency of our environment and society.

The 'industrial ecosystem' will not soon be complete because of our incomplete knowledge and inadequate technology, but it seems a reasonable direction to follow. The incentive for industry is clear: minimizing cost and staying competitive will be based on rational economic approaches that allow for internalizing external diseconomies. Equally clear are the benefits to society at large: technological progress will be achieved with minimum disruption to the environment and human health. By remembering that we are part of and embedded in the natural world, and that our well being depends upon its proper functioning, we may find the will to imitate the workings of the best natural systems and to construct an industrial ecosystem that has far fewer conflicts with the rest of nature.

ACKNOWLEDGEMENTS

The text is based on the article 'Strategies for manufacturing' by Robert A. Frosch and Nicholas E. Gallopoulos (copyright 1989 by Scientific American, Inc.). The authors are grateful to their colleagues Michael J. D'Aniello, Martin A. Ferman, Joseph Hunter, and Michael G. Wyzgoski, all of the General Motors Research Laboratories, for their invaluable help in preparing the original *Scientific American* article.

REFERENCES

Allenby, B.R. (1989) Industry and global climate change. In *The Enviromental Forum* 6, pp. 6–9.
American Iron and Steel Institute (1987) *Annual Statistical Report*, 1980–1987.
Anon. (1985) Flow of iron in the USA In *Modern casting*, pp. 36–37.
Anon., Why the uproar over incineration: *Modern Plastics*, June 1988, 66–67.
Anon., 1989*a* Solid waste concerns spur plastic recycling efforts *Chemical and Engineering News*, January 30, 1989, pp. 7–15.
Anon., 1989*b* Trashing a $150 billion business (1989) *Fortune*, August 28, 1989, 89–98.
Ayres, R.U. (1989) Industrial metabolism. In *Technology and environment* (ed. J.H. Ausubel and H.E. Sladovitch). Washington D.C.: National Academy Press.
Changes in the iron and steel inventory 1956–83. (1984) *Phoenix Quarterly* **16**, p. 8.
Combustion study shows no PVC link with toxicity. *Modern Plastics*, September 1987, 12–14.
Friedlander, S.K. (1989) Environmental issues: implications for engineering design and education. In *Technology and environment* (ed. J.H. Ausubell and H.E. Sladovitch). Washington D.C.: National Academy Press.
Gallopoulos, N.E. (1982) *U.S. automotive fuels – 1982 shapes a new outlook*. SAE Paper No. 821251.
Hunt, R.G. and Welch, R.O. (1974) *Resource and environmental profile analysis of plastics and nonplastics containers*. Kansas City, MO: Midwest Research Institute.

Lankford, W.T. Jr, Samways, N.L., Craven, R.F. and McGannon, H.E. (ed.) (1985) *The making, shaping and treating of steel*. Pittsburgh, PA: Herbick & Held.

Leaversuch, R.D. (1989) PVC recycling–landfill fodder? It doesn't have to be. *Modern Plastics*, March, 1989, 69–73.

Loebenstein, J.R. (1985) Platinum-group metals. In *Minerals facts and problems*. p. 595. U.S. Department of Interior, Bureau of Mines Bulletin No. 675. Washington, D.C.: U.S. Government Printing Office.

Nass, L.I. and Heiberger, C.A. (1986) *Encyclopedia of PVC* (vol. 1, 2nd edn). New York: Marcel Dekkar, Inc.

Odum, E.P. (1989) Input management of production systems. *Science, Wash.* January 13, 177–182.

Pearce, D., Markandya, A. and Barbier, E.B. (ed.) (1989) *Blueprint for a green economy*. London: Earth Scan Publications.

Peters, W. and Schilling, H.D. (1978) *Coal resources*. U.K.: IPC Science and Technology Press.

Petroleum reserves and U.S. consumption (1988) In *American Petroleum Institute Data Book*.

Rider, D.K. (1981) *Energy: hydrocarbon fuels and chemical resources*. New York: John Wiley and Sons.

Robson, G.G. and Smith, F.J. (1988) *Platinum 1988*. London: Johnson Mathey Public Limited Company.

Ruckelhaus, W.D. (1989) Toward a sustainable world. *Scient. Am.*, September 1989, 166–174.

The garage industry (1989) *The Economist*, April 8, 1989.

The World Almanac and Book of Facts (1987) (page 12).

U.S. Department of the Interior Mineral Commodity Summaries (1988) *World reserves, resources and U.S. consumption of minerals.*

United Nations (1984) 1982 *Statistical Yearbook*. New York: UN.

U.S. Department of Energy (1988) *International Energy Annual, 1987*. Energy Information Administration.

Worldwatch Institute (1987) *Mining urban wastes: the potential for recycling*.

Index